Engineering Design
Teaching Aids

Edited by Colin Ledsome

DESIGN COUNCIL
educational

ENGINEERING DESIGN
TEACHING AIDS

Published April 1987
in the United Kingdom by
The Design Council
28 Haymarket
London SW1Y 4SU

Typeset in Monophoto Bembo
and printed in the United Kingdom by
Jolly & Barber Ltd, Rugby

ISBN 085072 193 8

Foreword

In 1977 the Design Council's magazine *Engineering* published a supplement of educational articles as part of the Council's response to the Moulton report on engineering design education, which had appeared the previous year. The supplement was well received. Over the following three years further editions appeared under the *Engineering* title until, at the end of 1980, editorial responsibility was transferred to the Design Council's Education Section and *Engineering Design Education* was published as an independent magazine.

One of the features of the new magazine was an inserted booklet of educational resource material under the general title *Engineering Design Teaching Aids Programme* (EDTAP). The aim was not to provide a comprehensive or prescriptive collection of material but rather to demonstrate some of the ways engineering design resource material can be put together. The magazine is aimed mainly at tertiary academic staff, but its broad spread also makes it interesting throughout career education and training from late secondary through to industrial formation. The teaching aid programme is similarly written to serve as a basis for teaching at a variety of levels so that staff may develop material to fit their specific circumstances.

The early units were written speculatively in that no one knew what was needed and we fully expected to be told where we had gone wrong. However, the comments received have almost without exception been favourable and welcoming, so the style and content of the first units have been retained and developed and extended to new topics.

Requests for extra copies of several titles have suggested the need for a body of material and so the contents of the first 12 EDTAP booklets have been brought together in this volume as part of the Design Council's Teaching and Learning Resources programme. They are grouped in categories and additional material has been added and editorial corrections made at a few points.

The EDTAP booklets will continue to appear in *Engineering Design Education* and suggestions for new units or offers to contribute material in the same style are always welcome.

Colin Ledsome
Editor, *Engineering Design Education*
Spring 1987

The Design Council
28 Haymarket
London SW1Y 4SU

Contents

Introduction

Most of the units in the Engineering Design Teaching Aids Programme are the work of Colin Ledsome. Where this is not the case the source of the material, or the name of the author, is given with the unit concerned.

The following notes show the headings under which the units are arranged and the issues of *Engineering Design Education* in which they appeared.

Design Exercises

The exercises can be used in many ways from brief examples to illustrate a specific problem to full open-ended projects occupying significant student effort. The aim here is to state the basic need and perhaps provide some background information giving the lecturer the freedom to present the problem as appropriate.

DE1	Garden wheelbarrow	Spring 1981
DE2	Raising water by water-power	Autumn 1981
DE3	Design exercises in brief (20)	Spring 1982
DE4	Solar sail	Autumn 1982
DE5	Video loop cassette	Spring 1983
DE6	Lawnmower accessories	Autumn 1983
DE7	Wire strainer	Summer 1984
DE8	Coffee-time challenges 1	Autumn 1985
DE9	Coffee-time challenges 2	Spring 1986
DE10	Seeking a market	Autumn 1986

Design Investigations

To complement the design exercises these units take a simple market need and look at some of the many ways it has been met. The need may be for a specific product or a similar function within different products.

DI1	Sun-awning support	Spring 1981
DI2	Hinges	Autumn 1981
DI3	Can-openers	Autumn 1982
DI4	Folding structures	Spring 1983
DI5	Keyboards	Autumn 1983
DI6	Springs	Summer 1984
DI7	Light industry (Communicating by light)	Spring 1985
DI8	Cork extraction	Spring 1986
DI9	Quarter-turn actuators	Autumn 1986

Bibliographies

Design has only been generally accepted as a valid subject for engineers to study for a relatively short period. As a result the number of papers and books available is limited and rarely listed separately. These two lists were culled from an international listing by V Hubka of Heurista.

BY1	Books	Autumn 1981
BY2	Articles and papers	Spring 1982

Case Study

There is a great shortage of information recording the actual design process in an industrial environment. These brief case studies concentrate on specific details to demonstrate the design process in action.

CS1	APT power-car sole bar	Spring 1982
CS2	CAD on the floor	Autumn 1983
CS3	Doubles detect unit	Summer 1984
CS4	Design process 1	Autumn 1984
	Design process 2	Spring 1985
CS5	Clothes line	Autumn 1985

Design Analyses

An engineering designer often needs to model a problem mathematically in order to obtain quantitative data on which to base a design. This analysis frequently does not fall within the main academic subject areas for which ample theoretical work exists. This series aims to show the 'toolkit' aspects of engineering mathematics.

DA1	Tolerances	Autumn 1982
DA2	Weight control	Autumn 1986

Design Perspective

There are many aspects of engineering design which are only a small factor in any individual design but whose total contribution over a period of time is considerable: yet most are taken for granted. The wheel is a classic case. This series looks at some of the others.

DP1	Rope	Spring 1983

Design Management
Management is often perceived as a single
discipline based upon the problems of
business and personnel, but the problems of
managing the design process are very
different. Design is inherently unpredictable
and unquantifiable except in the broadest
terms, and therefore its management must
be adaptable and able to make decisions on
incomplete data. These units cover some of
the techniques and concepts which help the
design manager meet this challenge.

DM1 CPA: 1 Networks   Autumn 1984
DM2 CPA: 2 Float and resources   Spring 1985
DM3 CPA: 3 Networks in use   Autumn 1985
DM4 Design responsibility   Spring 1986

One other category in the same EDTAP series,
Learning Aids, has been published as
numbers 1–100 in the *Technical Files* series
from *Engineering* magazine. These are
available on microfiche from the Design
Council.

Design Exercises

Garden wheelbarrow

Requirement To design a wheelbarrow for use in a domestic garden, to carry normal loads over typical garden surfaces without excessive effort by the user or risk of damage to vulnerable surfaces such as lawns. The design should also be robust enough to withstand occasional overloads and mishandling. The barrow should be capable of being stored in the open with minimal weather protection. It should be designed for large-batch production and be competitively priced.

Specification The wheelbarrow should be able to carry normal loads of soil or stones up to a mass of 60kg, with an occasional overload with a mass of 75kg. The barrow should be designed to last at least ten years carrying 500 loads per year, 5% of which are overloads. The design should be stiff enough to avoid excessive bouncing under overload conditions, or sloshing of any semi-liquid loads such as wet concrete. The load container should be so shaped as to minimise spillage of heaped loads when moving up or down a 10% slope. The barrow should be capable of being pushed up a 30mm step without excessive effort and of standing on a 15% slope in any orientation with a full overload without toppling.

Basic design

Everyone has seen and used wheelbarrows, and there are even toy barrows for young children to play with. It may, therefore, seem trivial to look at the problems of designing one, and yet, as we will see, it is not as straightforward as it looks. A simple wheelbarrow has a single central wheel at the front. That supports the front end of a frame which carries an open-topped container for the load. When standing, the rear end of the frame is usually supported on a pair of legs. When moving, the rear end is supported by the operator using a pair of handles on the frame.

Configuration The first design consideration must be the overall configuration, the relative positions of the wheel, the load, the handles and the legs. What are the factors which determine these positions? Which parts of the requirement and specification are relevant? What do you know about using a wheelbarrow which will help make these decisions? Can you carry out any analysis work which would be useful here? Try configurations which are different from the usual design: could the load container be lower down or over the wheel, should it be designed for pulling instead of pushing? Do these other configurations have any particular advantages or disadvantages? Do not forget the human factors such as lifting the handles, steering the barrow, leaving room for necessary leg and body movements. (See the Ergonomics section that follows.)

Materials Before you can consider the details of the design and how all the parts fit together you must decide on the materials to make it from. The most likely ones are wood, steel, aluminium, glass fibre, and plastics. You may use one material or a number. Which parts of the requirement and specification apply here? Are different materials more suitable for one part than

another? You may find that some other design decisions have to be made before the materials are chosen. One major consideration must be the methods to be used when fastening and joining parts together.

Structure Apart from the wheel, a barrow is mainly a structure carrying loads. The load container is essentially a local load carrier to distribute the weight of the contents, be it sand or large stones, in such a way that it can be supported by the main frame. Should the container form part of the main structure or should it be independent? (This raises the possibility of a different barrow design with one frame and a series of interchangeable containers for different types of load. Did you consider that as a possible configuration?) In what forms are your chosen materials available and which would be most suitable for each part? Consider the details of shaping and joining the parts. What will be the order of assembly? Will there be adequate access to the joints at each stage? How will the parts be held as work progresses?

The wheel The diameter and width of the wheel, the shape of the tyre, and the type and position of its bearings are all points to consider carefully. What parts of the requirement and specification affect the wheel design? Should the axle be stationary with one bearing in the centre of the wheel or should the axle rotate with the wheel and have a bearing at each end? How will the shape and dimensions of the wheel affect steering and manoeuvrability?

Ergonomics The user of a wheelbarrow is closely concerned with its operation and various aspects of the human body and its capabilities must be considered. For example the height of the handles must be fixed so that the shortest person likely to use the barrow can push it without the legs of the barrow hitting the ground. Consider all the

ways that a user interacts with a barrow and how they affect the design.

Surface finish All of the materials require some consideration as to their surface finish. Resistance to weather, abrasion, and corrosion, ease of cleaning and appearance should all be considered here. The surface may be covered with paint or some other coating or simply left as it was formed, but even the latter should be the result of a specific decision and not simply an omission.

Commercial considerations The wheelbarrow is a product designed for the domestic market. Are there any particular features of the design which would make it more attractive to a potential customer and more pleasing to use? Can these be enhanced to make a better product from the customer's point of view?

Final considerations Having produced a design it should be carefully reviewed. Does it meet the original requirement and specification? During the design did you impose any extra requirements and have you met them? Were any early decisions superseded by later developments, but left in unnecessarily? What have you learnt from this basic exercise?

11

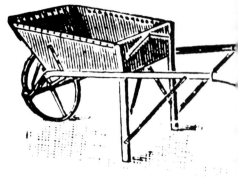

Further design work

Having produced an initial design, this may be used to provide a basis for the detail design work which in turn may suggest a review of the main design. The basic design and the work which went into it also provide a standard of comparison for looking at other designs more objectively.

Overall design It is often difficult to take an overall look at a design when you have been closely involved with the details, but many of the initial ideas were formed before you knew how they would affect the details. If you were starting again would you use the experience you gained in the initial design to change any basic concepts? Were there any features which did not receive enough attention? Did you consider, for example, the conflict between a deep load container which would reduce spillage and a low front for easy tipping? If a spadeful of earth were thrown in against the side of the empty barrow, would it be knocked over? If it were tipped up on end for storage, would the ends of the legs form a dangerous protrusion? Would it then be standing on its wheel, which would make it more likely to fall? If the main part of the barrow will just pass between two walls, will the user graze his or her knuckles? If the barrow is being pushed and one of its legs hits a solid object, is it likely to break off? Use your imagination to put your design in a number of working situations and see if it carries on working.

Now that you know something about barrows, ask yourself what changes you could make if you were designing it just to carry liquids or single large stones or to act as a trolley for a specific item such as a car engine in a factory.

Manufacture The details of the production process will depend on your design and especially on the materials used.

Wood Have you made the shapes of the component pieces as simple as possible to reduce the costs? Can they be easily cut, without excessive wastage, from standard stock sizes? Do the joints require the pieces to fit together exactly or can there be some variation in component size and still make the assembly? Are you using screws or nails and if so are they big enough and do any of them go into weak end-grain positions? If you are using metal joint pieces or any other mixed metal-and-wood constructions make sure any loads are adequately distributed in the wood/metal joints. Can components be fully finished in themselves or will some final trimming or hole drilling be needed on assembly?

Steel If you have any pressings or castings it may well be cheaper to redesign them to be made from standard sheet plate, tube or bar. If you require any bending or folding processes, check the standard handbooks for minimum bend radii and look at the corners where two bends meet. Is there adequate access for tools to fit any fasteners or carry out any welding? Check the handbooks for the edge distance of fasteners. The remarks

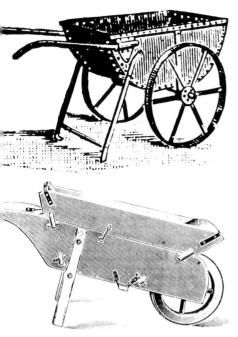

under wood about minimising wastage and dimensional tolerance also apply here.

Aluminium All the remarks about steel also apply here with an additional word of caution about corrosion. Aluminium by itself forms a surface oxide layer which resists further oxidation, but in contact with other metals it can accelerate electrolytic corrosion. Thus steel fasteners in aluminium must be plated or otherwise protected. Adjacent steel components must be suitably coated by a long-lasting material. It should also be noted that aluminium is not easily welded and joints of this type should only be used with caution.

Glass fibre and plastics Both of these require moulds which must themselves be designed and tested, which is expensive. However, the extra freedom to make more complex shapes and reduce joining problems can make it worthwhile. You may have evolved an all-plastics design or one with a glass-fibre load container on a tubular-steel frame. In all cases you should ensure that there are enough stiffening ribs to avoid excessive local or overall distortion. See if you can find a similarly-loaded plastics product to get some idea of how stiffening

ribs work. Make sure the material will be resilient enough to take the knocks of use without cracking. Remember that some plastics and resins deteriorate and become more brittle with age and exposure to sunlight, so the design must take account of the material properties near the end of its life and not just when it is new. All moulded components must be designed so that they can come out of the mould. This may sound simple, but if you look up a few references on plastics and glass fibre you may find design information on draught angles, shrinkage and other related factors. You may be using plastics or rubber for specific small components such as handle grips, bearing liners and tyres. In these cases you may well find that these components already exist or can be made from readily-available materials, eg bicycle handgrips, tube stock and existing industrial tyres. Catalogues or British Standards can be useful sources of this information.

Design comparison
With the knowledge and experience derived from your design work, you should now be able to look at other barrow designs and pick out their interesting features. Look in garden centres and department stores at the barrows on the market. If you have access to any barrows, try them out! See how they steer round sharp corners; are they stable on slopes when loaded; are the handgrips comfortable? If the barrow is a few years old, note which parts have deteriorated.

Another interesting comparison which can now be made is with historic designs. Visit an art gallery or library to look up paintings and prints which show street scenes or industrial activity. See how many different types of barrow you can find and how many details of design and construction you can pick out. Can you relate these to the historical content of the scene and activities illustrated? Has the artist put a design from his own era into a scene from earlier times?

Advanced design work

Having got this far there may not seem much more to do, and indeed for an ordinary garden barrow the design work you have done would probably be regarded as quite sufficient. However, for the purposes of this exercise the barrow design provides a convenient way of illustrating some of the detailed design-related activities which would normally only be carried out on more complex or critical products.

Structural analysis A wheelbarrow, of any reasonable design, has a fairly simple but not straightforward structure. It should be possible for you to sketch a structural model of your design showing its main structural components as struts, ties, beams, webs, flanges, etc, for both the standing and moving situations. Take the loadings from the specification and evolve a series of load cases including lateral and longitudinal loads to be imposed on your model. What safety factors should be included? Are there any other loadings which you think should be taken into account? Consider the various local loads which can occur on the load container. Now carry out a structural analysis of your model for all your static-load cases covering ultimate, proof, and fatigue cases as appropriate. An initial basic analysis using a simple beam model may be useful to give approximate figures to check your main analysis against. Check the stresses and buckling loads at the most heavily loaded points and at the key joints. In the light of your analysis, do you want to change your design? Are some parts not strong enough? Are any parts far too strong and could they be reduced in size without increasing production costs? Using a simple beam model (or the full model if the computing facilities are available), carry out a dynamic structural analysis to find the natural frequencies of the structure? Are these high enough to avoid being excited by a fast walking pace of the user? Do you need to stiffen the structure?

Joint design Choose a major joint in your design and examine the loads in detail in this area. If you have used fasteners, check the loads in each one for each load case. Check for shear, bearing, and tear-out failures. Are the fasteners ever in tension and if so are the bearing areas sufficient? Check welded or adhesive joints in the same way. Could a design modification move the joints to lower-stress areas?

Bearing design The wheel bearings are probably most heavily loaded when the barrow is fully loaded and angled over to turn a corner or tip sideways. Analyse the loads on the bearings under these conditions and use bearing-design data appropriate to the bearing type to carry out the detail design of the bearings. Check the way in which the bearings are retained. How will they be fitted during assembly? Can the bearings or the wheel and tyre be replaced when worn?

Suspension Some wheelbarrows include a simple suspension system with some springing between the wheel and the frame. This absorbs some of the shock loads and reduces load spillage due to jolting and also reduces user fatigue. A similar purpose is served by a softer tyre although this may result in excessive tyre wear. Investigate the design of such a system. Take the case of an overloaded barrow bumping up and down 30mm steps at normal walking speed. Devise a system to reduce shock loads to a reasonable level. Define 'reasonable' for yourself. What part does the stiffness of the structure play in this? Can the system still take the loads used in the bearing design?

Costing Costing exercises are notoriously unreliable when done by someone not directly connected with a production facility. However, for a simple wheelbarrow without complex, high-accuracy components the process is fairly simple. Carry out a costing exercise on your design,

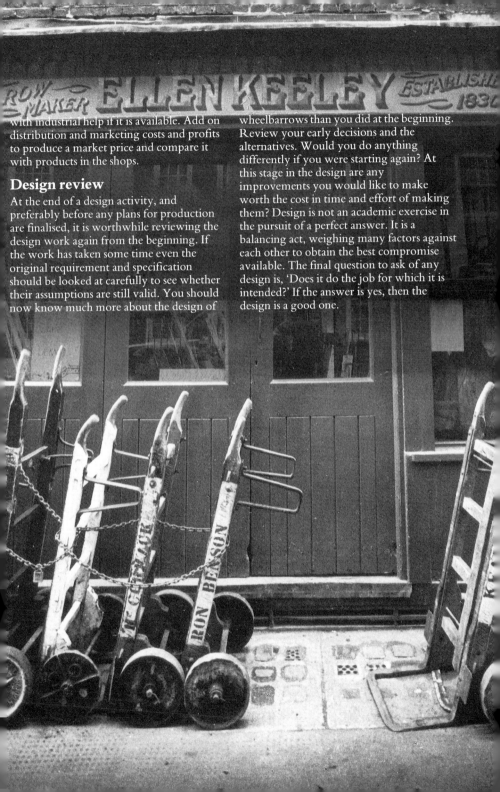

with industrial help if it is available. Add on distribution and marketing costs and profits to produce a market price and compare it with products in the shops.

Design review

At the end of a design activity, and preferably before any plans for production are finalised, it is worthwhile reviewing the design work again from the beginning. If the work has taken some time even the original requirement and specification should be looked at carefully to see whether their assumptions are still valid. You should now know much more about the design of wheelbarrows than you did at the beginning. Review your early decisions and the alternatives. Would you do anything differently if you were starting again? At this stage in the design are any improvements you would like to make worth the cost in time and effort of making them? Design is not an academic exercise in the pursuit of a perfect answer. It is a balancing act, weighing many factors against each other to obtain the best compromise available. The final question to ask of any design is, 'Does it do the job for which it is intended?' If the answer is yes, then the design is a good one.

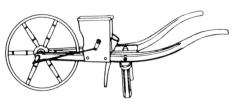

Teaching notes

A wheelbarrow is one of that large class of useful everyday items which has evolved over thousands of years into a clean simple product of man's ingenuity. The basic barrow performs its function well with an irreducible set of parts giving that intellectual satisfaction recognised as good design. The wheelbarrow was probably one of the earliest wheeled vehicles and its modern form can be traced back into early history. Even the details of its construction were fairly settled over a period of centuries until modern materials and production methods allowed a further phase of evolution to take place.

As a design exercise a wheelbarrow is particularly useful since it is a familiar, everyday object and requires no effort to understand its function or purpose. The three stages given here of basic, further and advanced design are roughly aimed at secondary up to O level, sixth form to A level and tertiary students respectively but this obviously depends on the individual circumstances. The format is aimed directly at the student and could be copied and distributed as it stands or used as part of your own presentation. It is intended as an aid, not as a course in its own right. Students taking on the more advanced parts of the exercise should first have completed the earlier parts if they are to fully understand the design processes.

The requirement and specification should be examined carefully by the students. Is it reasonable? Do any parts conflict? Do they want to add anything such as safety requirements or testing procedures? The consideration of configuration allows the introduction of force resolution and ergonomic factors. The ergonomic data can be obtained from references or investigated and derived by the students as a separate exercise. For guidance the normal palm height of an adult standing with arms straight down is about 780mm above the ground, the normal load which can be comfortably carried in that position is 400N, the distance between the handles should be about 500mm.

The basic decisions on choice of materials and overall construction are fundamental and students should have access to information on various types of material, including strength properties and the forms in which they are available. Information can be obtained from handbooks and standards or from the manufacturer's catalogues. (Caution: manufacturers may supply one copy of a catalogue but will be unhappy to receive a lot of requests from students at the same address with no prospects of sales.) The older wooden barrows frequently used the sides of the load container as part of the main structure as do the all-plastics or glass-fibre barrows, but those with metal frames usually have an independent or only partly-structural load container fastened to the frame.

Designing the wheel can almost be an exercise in its own right. The ability of the wheel to ride over bumps and soft ground is mainly dependent on the diameter of the wheel, which should be at least five times the height of the maximum step. Students

may like to investigate this for themselves. In a similar way the ergonomic aspects can generate a number of lines of investigation and opportunities for tests. There is a certain amount of scope for splitting a large group of students into a number of small teams to investigate these aspects and the alternative materials and construction questions, but they must co-operate to co-ordinate their activities and a project management team may prove useful.

The final considerations are important to the basic exercise in pulling the design together into a unified concept. If any drawings have been done or reports made, these should be in a presentable form. The student must review the exercise in order to discover what has been learnt from it and gain a better insight into the design process.

The 'Further design work' section is essentially a more-detailed look at the design and can be run as a continuation of the exercise without a specific break. Here more-detailed engineering drawings become an essential design tool so that each participant can understand the details under discussion. If computer-graphics facilities are available they may well be of use here. The details of joints and fasteners give a broad scope for design variety and it may well be necessary to limit the activity to a specific joint.

As a starting point for a historic investigation the wheelbarrow is particularly useful as it appears in many paintings and prints by artists such as Breugel, Hogarth or Agricola. In addition there are various types of barrow ranging

from the large two-wheeled costermonger's barrows, which have now almost disappeared from the streets of London but feature in songs and literature of Victorian times, and the rickshaws of eastern countries, to the simple wheeled plough and the supermarket trolley.

The advanced work certainly introduces concepts not normally applied to wheelbarrows, but they are valid nevertheless in this context because they provide a simple basis for more complex ideas. The details of the exercise at this level will be dependent on the course being followed, the expertise of the students, and the points which the lecturer wishes to make. The exercise given here is therefore much more a suggestion of the possibilities available at this level than a complete coverage of the field.

Further reading

Engineering Design Guides, published by Oxford University Press:
01 *Introduction to fastening systems*
02 *Adhesive bonding*
06 *Welding processes*
12 *Metal forming II: pressing and related processes*
17 *The engineering properties of plastics*
21 *Metal corrosion*
24 *Plastics mouldings*
29 *The selection of materials*
Culpin, Claude, *Farm machinery*.
Brichton and Sharp, *From project to production*.
Also check relevant British Standards.

Raising water by water power

Requirement To design a device which will use the power available from water falling through a small height to raise a smaller volume of water through a large height. The design may be for one of two configurations: one where the device is raising water from a deep shaft and the other where water is being lifted up a hill to a higher level. In the first the water power is available at the top of the lift, in the second it is at the bottom.

Specification In both cases the height of lift of the water should be 100m at a minimum rate of flow of 90 litre/min and the power source is a 10m fall of water with a dependable flow rate of 1500 litre/min. For the shaft option the water should be raised to the same level as the outflow from the device. For the hill the water should be derived from the same reservoir which powers the device.

Basic design

These are the classic engineering design problems of the early industrial era. Water had to be drained from mines so that new supplies of minerals could be exploited; water had to be raised to irrigate high farmland and later to fill canals and supply drinking water to growing hill towns. Before the coming of the steam engine the only dependable source of power was flowing water. Even today there are many areas of the world where this form of device is still the most efficient and dependable choice, particularly for water supplies. Most mining operations are now carried out using modern pumps and other power sources.

In our situation any external power source is discounted and the problem is the same as that confronting the early engineers except that our solutions can now use modern materials and techniques and the benefits of hindsight. As in any design work the basic design problem is that of conceiving an overall configuration for the device which might satisfy the requirement, and then refining it until a careful analysis shows that it will meet the performance specified. Then, and only then, can the details of the design be investigated.

There is broad scope in this open-ended exercise for updating the many devices which can be found in historic literature as well as conceiving new designs to solve old problems. The possible solutions fall into a number of categories:

Hydraulic or hydro-pneumatic Here the device is a system of pipes, valves, and vessels with perhaps a few mechanical links to provide a control system.

Hydro-mechanical In this the falling water drives a mechanical device which conveys or pumps the water to be raised.

Hydro-electric Strictly hydro-mechanical-electric, where the falling water drives a mechanical system which turns the shaft of a generator which in turn provides the current to drive an electric-motor-driven pump or other mechanical device to raise the water. The likely losses of energy at each conversion step would accumulate to make this type of solution low in efficiency and it may not be possible to reach the required performance figures.

As a basic design exercise simply conceive and sketch or otherwise present the overall configuration of a device to satisfy one of the two requirements. Then examine in more detail the principles of operation of your device and derive the governing theory necessary for its analysis.

Further design work

Carry out a detailed design study on one of the key moving parts in your device. Look at materials, configuration, assembly and maintenance, corrosion, and lubrication. What other factors should you consider?

Using the theory derived earlier carry out an analysis of your device to determine its efficiency in terms of work done on the water raised to power supplied by the falling water. Examine the water pressures which occur in various parts of the system and carry out a simple structural analysis to find a safe wall thickness for your pipework (include a safety factor). You must first, of course, find the pipe diameter you need to take the expected flow. Whichever of the alternatives you chose for your first design now try to apply the same principle to the other. Although the two problems are very similar in many respects, their solutions may not be.

Advanced design work

You should now have at least an outline solution to both versions of the problem. Now suppose that we turn the requirement round and have a high pressure but low

flow rate water supply from an elevated reservoir which must be used to lift a large volume of water through a small height. Assume only one case where the water must be lifted from the level of the high-volume supply.

New specification The power source comes from a reservoir 120m above the device with a maximum flow of 20 litre/min and water must be raised 5m at a minimum rate of 250 litre/min.

Using the experience you have gained on the earlier designs or trying a completely new approach, design a device to meet the new specification.

Teaching notes

Since this is an open-ended exercise the information provided for the students is restricted to avoid channelling their work in a particular direction. However, this implies that the lecturer or teacher leading the class must have a broad understanding of the many possible ways in which solutions can develop and enough detailed knowledge to be able to give help and advice when it is needed. These notes cover a broad range of design concepts which students may stumble upon or, if desired, they may be used as guidelines to act as an ideas source. The exercise may be used to form the basis of a full investigation of this type of device. Most large encyclopaedias will provide basic information on a number of them and a few references are given here for further information. There is also wide scope for using the historic background of these devices to investigate other forms of water powered devices, pumps, mining and agricultural and industrial machinery.

As indicated in the exercise the solutions fall into three categories, of which the hydro-electric ones are hardly viable in this context. For students with little experience of hydraulic or hydro-pneumatic devices, the most likely direction of thought will be

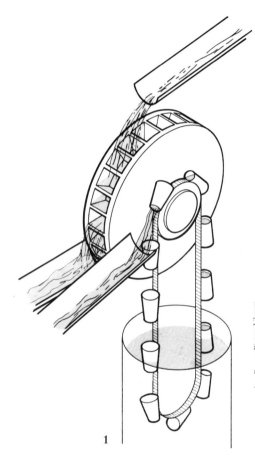

1

2

towards mechanical devices. The power available from the falling water must be converted into mechanical movement, either via a water-wheel, a turbine, or possibly a reciprocating mechanism. This motion is then used to raise water with a pump or some form of conveyor or a series of these. The most likely is the water-wheel.

W ater-wheels are usually classified according to the level at which the incoming water reaches the wheel relative to the wheel centre. Undershot wheels were known in

the fourth century BC and were widely used for milling in the Roman Empire. The overshot wheel was being developed in Byzantium in the fifth century AD. In this country the water-mill is first recorded in AD 762 at Dover. By the Domesday Survey of 1086 some 5624 were listed south of the rivers Trent and Severn. One of the earliest devices which will serve our purposes was the bucket chain. A series of containers on a continuous loop of chain is carried round by the power of a water-wheel. This will serve to lift water from a shaft provided that the mechanical advantage conveyed by the relative water-wheel to chain-wheel radii is

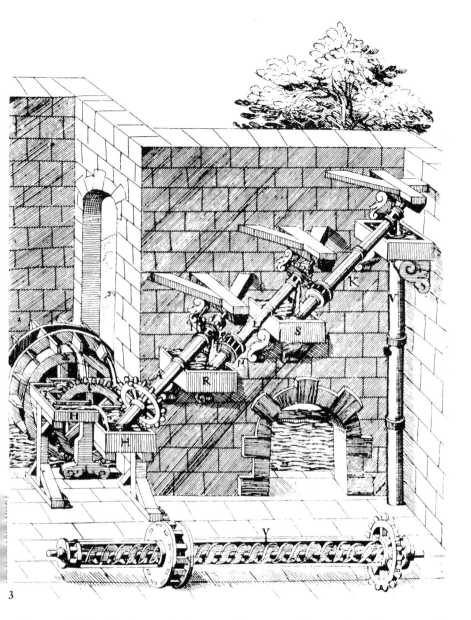

3

sufficient. This is the simplest mechanism we can consider (figure **1**).

Another mechanical water-lift with a long history is the Archimedean screw. This is a long cylindrical pipe with an internal spiral plate. When placed on an incline with its lower end in water, rotation of the cylinder will raise water to the upper end. To achieve the height of lift we want would require either a very long cylinder or a series of

short ones with a complex drive system. It is unlikely that the efficiency of such a device would be high enough to achieve the desired flow rate.

In 1556 Georg Bauer, better known as Agricola, published a 12-volume survey of methods of mining and working metals. His records of mine machinery include banks of water-pumps driven by water-power. As figure **2** shows, these were simple single-acting pumps, each lifting water a height of perhaps three or four metres, acting in tandem to give the required height. This system gradually evolved over the centuries until the latter part of the nineteenth when giant wheels such as the 22m-diameter wheel at Laxey in the Isle of Man (figure **4**) were erected. This is still in existence although the mine it served is derelict. It developed 185hp (138kW) and lifted $1.137m^3/min$ through a height of 457m. Although the pumps in the shaft were single acting and worked simultaneously the work load was smoothed somewhat by a counterweight which was lifted during the non-working half of the wheel's cycle and added to its energy during pumping.

In parallel with the mine drainage systems similar devices were being developed for irrigation and water supply. Here a series of pumps require a more complex mechanism than in a mine, to avoid having the main drive links in compression, so the concept

4

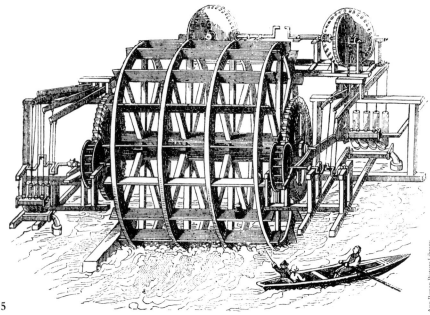

5

of a number of pumps in parallel was developed. This configuration smoothed out the loading on the wheel and allowed the necessarily higher pressures to be developed. A device of this type (figure **5**) was installed between two of the arches of London Bridge from 1581 until 1822. It floated up and down with the tide and used the fast flowing water between the arches to power the city's water supplies.

Pumps developed from simple lift pumps, with water above the piston, to single acting (figure **6**), then double acting pumps. Their drive mechanisms grew from simple cranks to double- and triple-throw cranks and moved on to eccentrics before their source of power changed from water to steam and gas. Any of these changes provide useful topics for more detailed investigations beyond the scope of this exercise.

Pumping water to a higher level took on a new aspect in 1775 when John Whitehurst gave details of the first hydraulic ram to the Royal Society. The original ram required an operator periodically to close a tap on the outlet from the supply pipe. The resulting sudden increase in pressure was sufficient to open a second valve into an air chamber. This air acted as a spring to absorb some of the pulse of energy and partly smooth the flow of water out through a delivery pipe (see figure **7**).

In 1797 Joseph Michel de Montgolfier, co-inventor of the hot air balloon, improved on this by introducing a self-acting valve in place of the tap, thus removing the need for an operator. The whole device was redesigned by John Blake in 1868 and patented as the Hydram (see figure **8**).

This device has been used successfully in many parts of the world and is still available in a range of sizes from John Blake Ltd. Cases are on record where a Hydram has worked steadily for over 60 years pumping thousands of litres per day using a supply head of only a few metres.

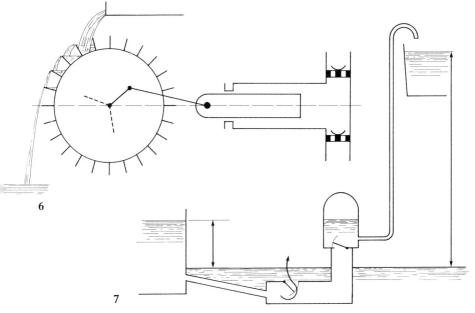

6

7

In the Hydram water is allowed to flow with considerable force and speed down an inclined drive pipe into the Hydram. The water spills out through the waste valve until the pressure on the rubber disc in the valve causes it to close. The water flows on through the pump body – still with speed and force – through the delivery valve and into the air vessel. The air within the vessel – acting much like a spring – absorbs the momentum of the water until the characteristic 'soft thump' noise of the Hydram is heard when the pressures reverse in the form of a shock wave. This travels back through the water, out through the ram and drive pipe when the waste valve reopens and the delivery valve closes. The water trapped in the air vessel is pushed out through the delivery pipe and into the storage tank. The complete cycle then repeats once again. The action being continuous, some of the water is always flowing through the delivery pipe.

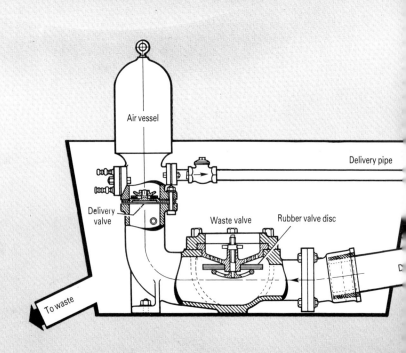

Air vessel

Delivery pipe

Delivery valve

Waste valve

Rubber valve disc

To waste

8

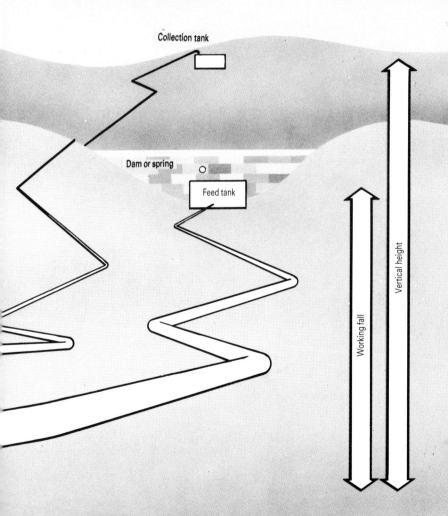

Collection tank

Dam or spring

Feed tank

Working fall

Vertical height

25

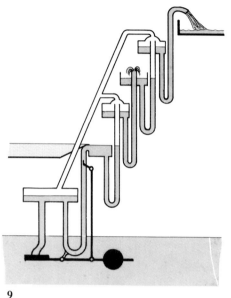

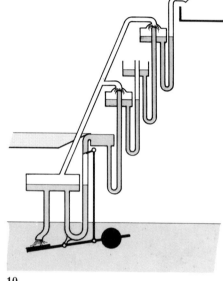

9   10

Another all-hydraulic system proposed in the 1920s is the Hydrautomat which uses a system of pipes and chambers to produce the same effect as a bank of pumps in series. Falling water from the power source alternately fills and empties a large vessel raising and lowering the pressure in a pipeline above and below atmospheric pressure. This in turn raises and lowers the pressure in a series of closed vessels ascending the hill. Thus water is alternately raised from open vessels below them and pushed up to open vessels above them. This chain of open and closed vessels forms the hydraulic equivalent of a series of mechanical pumps, thus raising water to the top of the hill. It is not known if any of these devices were built.

There is ample scope in this exercise for branching off into more detailed investigations of water-wheels, pumps, valves, mechanisms for converting rotary to reciprocating motion, seals, and other topics. The additional exercise of the high-pressure, small-flow-rate supply being used to power a low-head, large-volume flow introduces extra possibilities such as Pelton wheels, hydro-electric systems, and all-hydraulic solutions where the momentum of a small jet is imparted to a large flow.

Further reading
Armytage, W H C, *A Social History of Engineering*, Faber.
Starmer, C, 'Blake's Hydram', in *Chartered Mechanical Engineer*, March 1981.
Burstall, A F, *A History of Mechanical Engineering*, 1970.
Landels, J G, *Engineering in the Ancient World*, 1978.
Vallentine, H R, *Water in the Service of Man*.
Isle of Man Tourist Board, *Lady Isabella – The World's Greatest Water Wheel*.

Design exercises in brief

Some design exercises do not lend themselves to lengthy specifications and others, particularly the Design-Build-Test type, are dependent on locally available materials and facilities for their detailed requirements. In most cases a very brief description is all that is necessary to provide the basic idea to be turned into an exercise appropriate to the local situation. Here is a listing of some of these types of exercise. Some have been collected from universities, polytechnics and colleges, often from more than one source, others are suggestions which have not been tried as far as we know.

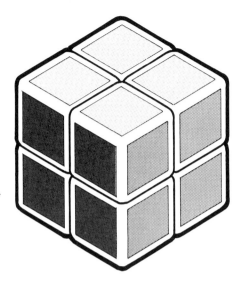

1 Design a kitchen tool which will slice a ripe tomato as easily as an egg-slicer cuts a boiled egg.

2 Devise a Rubik Cube mechanism for a 2 × 2 × 2 cube. The normal cube puzzle is a 3 × 3 × 3 which can be rotated on any of the six dividing planes without falling apart. The problem here is to devise an equivalent mechanism for the simpler 2 × 2 × 2 cube.

3 Design a small self-steering vehicle which will follow a prescribed path such as a sinusoidal curve. Such a vehicle may be powered by an elastic band, battery-powered motor, clockwork, or similar power source. Do not use a microchip!

4 Devise a microprocessor-controlled steering mechanism for an electric lawnmower which can be steered round the edge of a lawn once and will then deduce its own path to mow the lawn, avoiding its power cable, and remember the route for later use. The system should make some allowance for emptying the grass-box.

5 Design a spacecraft docking mechanism. Spacecraft are steered towards each other until their docking systems make contact. These mechanisms must initially latch together and then move to correct any misalignments between the vehicles. They pull the two vehicles together and finally make a structural clamp. All of this must happen with room for airlock hatches and seals, electrical and pipe connections within the connected area. Any two vehicles should be able to dock with each other so their docking parts should have identical mating features.

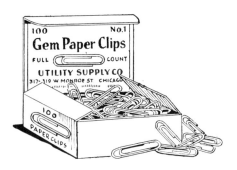

main difficulty here is to make the end of the inner loop lie at a slight angle to the main plane of the clip.

10 The feedback problems of public address systems can be considerably reduced by changing the frequency of the output by a small amount. A constant addition or subtraction of about 7Hz gives the best feedback reduction with little appreciable distortion of speech. The effect on music is not usually acceptable. Design the necessary circuits to achieve this frequency change.

6 Design a mechanism which will plait or braid three cords into a rope.

7 Devise a folding umbrella which will fit into a coat pocket when folded, about the size of a thick paperback book, yet will provide adequate protection from the rain and withstand moderate gusts of wind.

8 Devise a coin-sorting machine to sort a mixture of coins in a hopper into denominations.

9 Design a machine to manufacture paper-clips from wire supplied on a drum. The

11 Design a simple pencil-sharpener which will form a chisel edge.

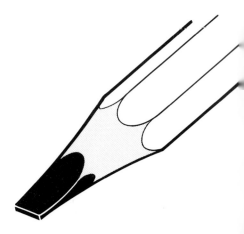

12 Devise a control circuit to change the speed of a small synchronous motor by varying the phase of its power supply.

13 Design a simple tool to notch floor joists to accept pipework. When installing new water or gas pipes under an existing floor, the joists must be notched out to fit the pipes. Design a tool to do this quickly and easily without an external power source.

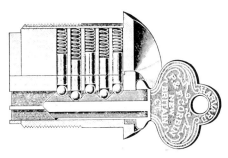

14 Design a simple tool for the home-brewing market for fitting crown-type metal caps on to beer bottles.

15 Carry out a feasibility study on a spiral escalator to fit in a vertical shaft, replacing a lift system at an underground station.

16 Design a tent to be used as a temporary service bay for cross-country vehicles. There must be a winching point in the centre of the roof capable of supporting a one-tonne winch and sufficient vertical clearance to lift out an engine.

17 Design a lock system where each person only needs to carry one key and the locks may be set to accept a number of predetermined keys. The locks may be mechanical or use an electronic system to read a plastic-card key. The system should not take up much more room or be much more expensive than any normal system it may replace.

18 Empty tin cans represent an unused source of sheet steel and aluminium. Design a set of tools for the DIY market to cut out and flatten the material then to cut, bend, and press it into useful products. There is scope here for a brief research programme into the formability of these materials.

19 Design a gyroscopic system to detect the rotation of the earth and use it to regulate a high-accuracy clock mechanism.

20 Design a device to support a curtain and allow it to be drawn across a window without needing a fixed rail across the window frame. The device should be easily fastened to a normal window frame, be completely concealed by the curtain and provide adequate support for the curtain in normal domestic use.

Solar sail

Requirement To design a simple spacecraft, powered by the photon pressure of solar radiation on large sails, to compete in a race from low earth orbit to the vicinity of the moon. Such a race is expected to be staged in the last years of this decade or the early 1990s. The design conveniently splits into several sections which may be considered independently.

Specification Solar photon pressure in the vicinity of the earth averages 4.65×10^{-6} Nm^{-2} ie very small. The launch package should fit into a volume 2m in diameter and 500mm deep. Details of a possible launch configuration are shown in figure **1**. Launch mass should not exceed 200kg and the package should be designed to withstand launch loads and vibration levels (not

specified here). The competing spacecraft will be put into an elliptical earth orbit $36,000 \times 3000$km. Using only the photon pressure of solar radiation for propulsion, the first craft to extend its apogee to lunar orbit and go behind the moon would be the winner.

Structural design

The structural problem is one of providing support for a large area of thin sail material with a light structure which will unfold

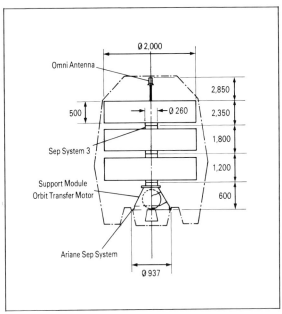

1 *A possible configuration for a triple solar sail launch by Ariane*

from a small package. The sail material will probably be a film of Kapton 2.5μm thick having a mass of about 4.8g/m². Since the centre of mass of the craft must lie on the line of the total pressure load, the configuration will probably consist of a control package with radiating spars supporting a polygon of sail, but you may wish to investigate alternatives.

The performance requirements dictate that the design should have the maximum-possible sail area. A craft with an area-to-mass ratio of 14 would take over two years to finish the race, an area-to-mass ratio of 25 reduces it to about a year but the problems of deploying and supporting 5000m² of sail are formidable.

Various solutions have already been proposed. Complex folded trusses carry a considerable weight penalty in joint pins and deployment springs and are likely to prove too heavy. Telescoping structures would need so many stages that the overlapping sections and locks would again make a heavy structure and reduce the reliability of the deployment. One solution uses flexible spars, rather like giant fishing rods, wrapped on to a drum. Once released they uncoil and straighten out carrying the folded sail with them (figure 2). Remember that the loads on the sail are very low and so the deployed structure need not be substantial. Spars of this sort would be braced in the plane of the sail by reinforcement in the sail itself. Out of that plane the spars may be deep enough in themselves or may require bracing with a

guy system to a central mast or other feature.

The most severe loads on the structure will arise during manoeuvres and will depend on the steering system used and the race strategy (see below). It is important to appreciate the space environment. The lack of gravity means that there is no weight to

2 *Test deployment of half-size sail by the World Space Foundation, August 1981*

support, but also there is no weight to hold things down or pull things straight. A folding structure has kinetic energy as it deploys, which must be absorbed when it reaches its desired position or it will keep on moving and tear a sail or set up vibrations which wreck the whole craft. No air means no aerodynamic loads, however it also means that there are no residual air molecules and sliding surfaces may weld together. Thus it may be possible to unfold the structure soon after leaving the atmosphere, but you will not be able to rely on unlubricated bearing surfaces for steering systems.

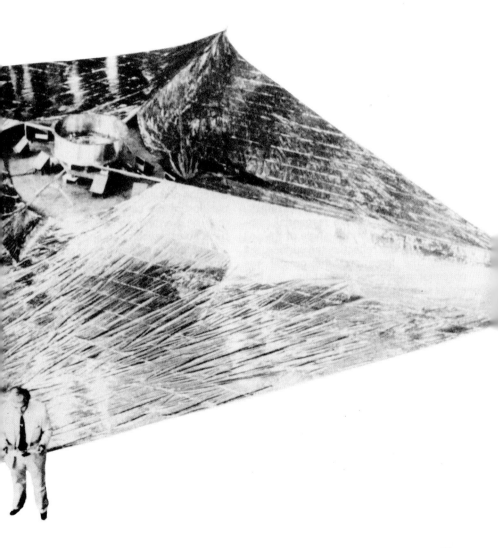

Manoeuvring system Sailing in space is not the same as sailing on the sea. A sea craft uses both the air and the water by balancing the available air pressure from the wind against the ability of the hull to move much more easily through the water longitudinally than laterally. This allows a boat to move at an angle to the wind and even tack into the wind. In space there can be no tacking and only a limited ability to move across the direction of solar pressure. The photon pressure referred to here is radiation pressure and not the solar wind of particles emanating from the sun. These particles would pass straight through the thin material of a solar sail. The direction of the radiation pressure depends on the reflectivity of the sail surface. Absorbed radiation applies a pressure in the direction directly away from the sun whereas reflected radiation may have a tangential component depending on the angle of the sail (figure **3**). The more reflective the surface, the greater the force and the greater the potential manoeuvring capability. But it is not as simple as that.

A boat on the sea can use a rudder to maintain the angle between the sails and the wind. In space you have no rudder and an angular position can only be held if the resultant pressure force passes through the craft's centre of mass in a stable configuration, ie a small rotation moves the craft in such a way that the movement is opposed. Even so some form of steering is needed to avoid oscillations. No atmosphere means no damping by air resistance, and so even a basically stable configuration needs an active steering system to make maximum use of the available pressure. In addition to maintaining a required angle, the manoeuvring system must be able to change the angle at least twice during each orbit to achieve acceleration while moving away from the sun and avoid deceleration while moving towards it.

Manoeuvring can be effected only by changing the relationship between the line

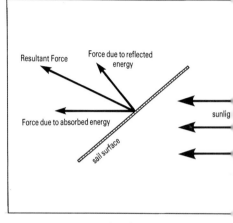

3 *Forces acting on a sail surface*

of action of the total pressure force and the position of the centre of mass. The sails may be warped to change their shape, have holes which open and close, or have small steering sails which rotate or unfurl to change the pressure pattern. Alternatively, the sail may be fixed and a steering mass moved about. Both approaches have their problems. Perhaps a major decision to be taken at an early stage is whether to have a single-sided or a double-sided sail. Figure **4** shows a typical manoeuvre sequence in orbit which meets all the acceleration requirements but turns the craft through 180° in each orbit thus requiring a configuration which is stable and manoeuvrable in two orientations.

Control system Because of the multitude of factors which could affect the race and the way in which the race strategy must change as the race proceeds it is not possible to program a small automatic system to cover all the eventualities. Therefore it is necessary to devise a control system which will carry out a looped sequence of simple manoeuvres once per orbit yet may have that sequence modified on instruction or even allow direct control from the ground. The amount of autonomy which could be

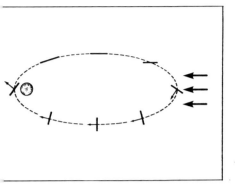

4 *The orbital manoeuvres of a moon racer.
Energy is picked up on the inward leg of the
orbit; while on the outward section the sail lies
parallel to the sun's rays to prevent deceleration*

given to the craft would depend to a large
extent on its on-board position sensors. Two
direction sensors will suffice to fix its
orientation. One fixed on the sun would
seem the best choice, but the other might
use the earth, the moon, or a distant star as
its reference. The ease of reacquisition after
an earth-shadow tumble (see below) might
dictate the type of sensor. The angle between
the earth and sun directions would give the
position in orbit to correlate with orbital
parameters acquired by ground tracking
stations.

Ground tracking should not be too
difficult since the large sails will be clearly
visible on any reasonably cloud-free night
for optical measurement. This may be used
to produce overall strategies for orbital
manoeuvres leaving the on-board system
to sort out the immediate tactics to meet
the strategy requirements. A small
microprocessor and solid-state memory
controlling a few motors or other actuators
should suffice for hardware along with the
necessary power system, with solar cells, and
the telemetry equipment. Designing the
software may well be more difficult if it is
to give all the necessary intelligence to the
hardware 'crew'.

Race strategy The race strategy directly
affects many of the design parameters and so
must be considered as part of the design in
the same way that software requirements
affect the choice of microprocessors. During
the early part of the race, orbits will be of
short duration and will be within the earth's
shadow for a significant time. If a small
steering manoeuvre is under way when the
craft enters the shadow, it will lose all
steering capability and simply tumble until
it re-enters the sunlight. The craft must then
rapidly re-acquire its position datums and
take appropriate action to stabilise its
orientation. The speed with which this can
be accomplished would have a considerable
effect on race performance.

As the race progresses the effects of being
near the earth reduce, the time in full shadow
reduces, gravity gradient effects become
insignificant, and residual atmospheric drag
becomes negligible. However, the craft
becomes less accessible from the ground as it
passes across the sky less and less frequently.
It may be necessary deliberately to use a less-
than-optimum race strategy to ensure that
the near-geostationary period occurs within
sight of a ground station or corrective action
may not be possible for a period of months.

During this period the apparent motion of
the craft across the sky will change from a
west to east movement to a near stationary
position in the sky then to an east to west
motion. The increasing eccentricity of the
orbit will mean that the apparent motion
will also become more erratic and so
ground-based observations must be
converted to real motion before being used.

For the later part of the race most of the
earlier problems disappear, but new ones
take their place. Within the earth-moon
system there are a number of areas in the
combined gravitational field where changes
in field strength and direction could pull the
craft off course. These areas could be avoided
or used to advantage to make the outcome
of the race uncertain right to the end.

Teaching notes

Kepler, in the sixteenth century, postulated that the energy pouring from the sun exerted a force on objects in its path. James Clerk-Maxwell refined the idea in the 1870s by proposing that electromagnetic radiation exerted a 'small but finite' pressure on an illuminated surface. It was another 30 years before Lebedev in Russia and Nichols and Hull in the United States demonstrated this effect experimentally. In 1924 Tsander and later Tsiolkovsky suggested using this pressure for propulsion, but little more happened until Garwin re-examined the idea in 1958 and coined the term 'solar sail' and Kraft Ehricke carried out some studies two years later. The concept surfaced in a few science fiction stories, notably *Sunjammer* by Arthur C Clark, which involved a race to the moon with manned craft. Then in the early 1970s solar sails were seriously considered by NASA for a probe to Halley's Comet in 1985/86. Solar sails or electric propulsion could have carried larger payloads than present-day chemical rockets on such a mission, but financial restraints ended NASA's plans for a 624,000m² sail.

At the same time the European Space Agency were interested in a small demonstration flight using an Ariane-launched sail. This was abandoned by ESA, but has been taken up by various private groups of engineers in Europe and the United States and it is on their work that the idea for this exercise is based. Current work indicates that such a race could take place towards the end of this decade. Since virtually the entire population of the earth would be able to see these craft on any reasonably clear night for two or three years, the interest in such a race would be high and the attractions to sponsors of individual craft and the contest itself should ensure that the relatively modest finance will be forthcoming.

There is little in the way of high technology involved in this exercise except

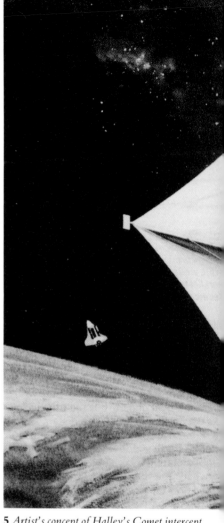

5 *Artist's concept of Halley's Comet intercept probe.* © NASA

perhaps for the materials used. The successful craft will be the one which exhibits the maximum design ingenuity to obtain a high sail-area/mass ratio and then is controlled well enough at each stage of the race. The designer has to have an understanding of the space environment and the ways in which it varies from the familiar earth surface conditions.

A lack of gravity is a more fundamental factor than many realise. The ground provides a reaction base for forces and hence allows us to control translation and

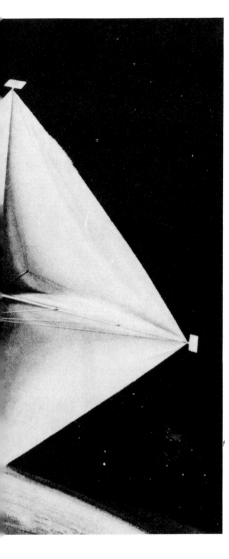

long or short in comparison to the other two, this produces a pendulum effect such that a long axis will tend to stabilise in a radial position relative to the earth and a short axis will tend to become tangential. Any manoeuvring system must exert forces strong enough to overcome this effect or make use of it. In a similar way, any long conducting path on the craft moving through the earth's magnetic field can have a current induced which in turn generates its own field which exerts a turning effect on the craft. This problem can be virtually eliminated by incorporating insulating material in the structure to reduce the lengths of conducting paths.

Micrometeorites and other particles should have little effect on the craft if the control system has some impact shielding. Particles hitting the sail would simply pass through leaving a small hole. This type of damage will gradually reduce the effectiveness of the sail as the race proceeds, but it should only delay the closing stages of the race by a small amount.

The gravitational field anomalies referred to in the Race Strategy section are of course around the Lagrange points. The first of these likely to be encountered is the one directly between the earth and moon. This one is not stable and craft approaching it would be deflected away. This may be used to advantage and may have a considerable effect since it will be encountered a number of times in the later parts of the race. The other three anomalies of interest all lie on the moon's orbit, one directly opposite the moon, which is also unstable, and the two Trojan positions 60° ahead and behind the moon, which are stable. These three will only affect the craft in the final weeks of the race and when careful use of these points could win a close-run race.

ACKNOWLEDGEMENT The information on which this exercise was based was obtained from meetings and papers of the British Interplanetary Society, 27/29 South Lambeth Road, London SW8 1SZ.

rotational motion. In space there is no ground and so all forces must be reacted by the craft itself; there are no insignificant forces. Newton's Laws of Motion are paramount. There is a gravitational field but its overall effects are exactly balanced by the centrifugal effects of the orbital motion. The word 'overall' is important here because the field gets weaker as you move away from the earth and so the centre of gravity of the craft is slightly closer to the earth than the centre of mass. On a compact satellite this has no effect, but if one axis is particularly

Video loop cassette

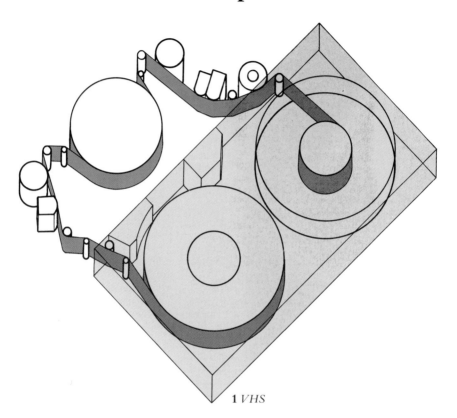

1 *VHS*

Requirement Many display, exhibition, and educational uses of video cassette recorders need repeated playback facilities. Currently this is only convenient with rather expensive machines which repeatedly rewind and replay sections of tape. This is not completely satisfactory, since there are long gaps between programmes and the constant stop–start sequences produce excessive wear. This exercise involves the design of a tape loop cartridge which will fit in the standard cassette envelope of a video recorder giving repeats of a recorded programme without excessive gaps on a much cheaper machine. Such a loop system could also be used for instant replay where long-term recordings were not required.

Specification The cartridge must be designed to match one of the standard formats: VHS, U-Matic, Betamax. The tape must load in the usual way, figure **1**, and the various drives and lock connections must be accommodated in such a way that no undue strain is put on the recorder mechanism. Cassette loop times will depend upon the amount of tape which can be accommodated within the device, but the design should allow a range of cassettes to be produced each with a different repeat time. A range from 30 seconds to 5 minutes in 30-second increments is suggested as a target range, but if longer times could be accommodated there could well be a market for them.

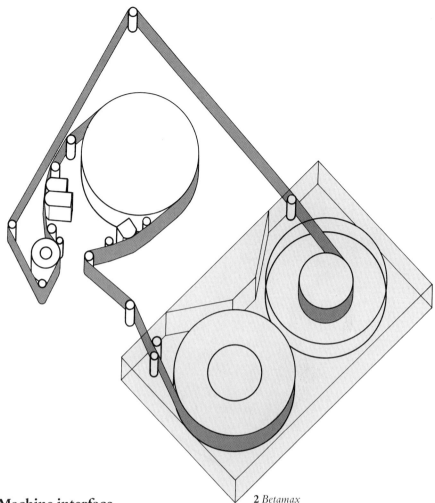

2 *Betamax*

Machine interface

Considerable interface information can be gained from examining a standard cassette, but care should be exercised and in particular the tape should not be handled if the cassette is to be used again. Any internal examination of a recorder should be carried out only under expert guidance, but judicious observation of a machine in action may well provide sufficient data for this exercise. Cassettes have a tape cover which is unlocked and lifted when it is inserted in the player. There is also a reel lock which is disengaged at the same time. The reel drives serve two purposes: the simple fast-wind fast-rewind function, with direct drive on one or other reel, and the tape-tensioning function, where a small torque is maintained on both reels to aid the normal tape drive and reel in slack tape. For very short loops there may be little need for fast winding, but if any editing or dubbing is to be done when recording the programme, fast winding becomes essential via the reel drive. Similarly, that part of the tape passing through the record/playback mechanism must not be slack and tension must be maintained. If the reel drives are not used they must be restrained from spinning due to machine torque.

Tape storage

There already are specialised tape loop machines used in professional studios which simply consist of a vertically mounted

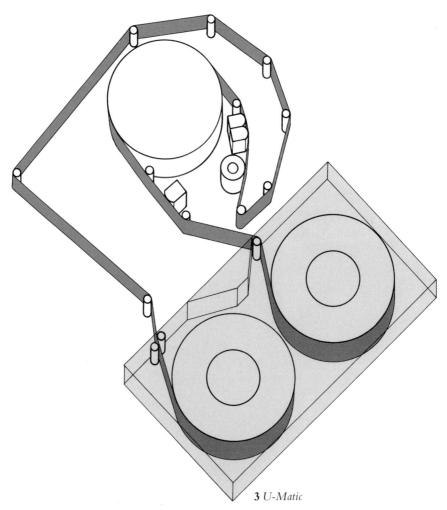

3 *U-Matic*

playing mechanism positioned above a wedge-shaped storage container. The loop of tape falls loosely into the store in coils and is drawn out again without tangling. In the device proposed here the limited space precludes this loose solution, except perhaps for the shortest loops. Hence storage must use a system of rollers or guide pins with some allowance for the amount of tape threaded into the machine upon loading the cassette.

Teaching notes

Recording loops of the type described in this exercise have a number of other uses. Many tests of observation, psychological tests, teaching aids, and so on use repeated viewings of material. For security purposes a continuously updated recording of the last five minutes might identify a bank raider or shoplifter without accumulating hours of unnecessary records or leaving rewind gaps in the recording. A supplementary exercise might explore this market.

The design will depend on the tape format chosen but the main factor is one of space available.

The main problem to be solved is that of tape storage in a continuously available form which can be fast wound. The tape may spiral or zigzag, but careful positioning of the guide pins or rollers is needed to get the maximum loop in. This exercise is mainly a test of ingenuity.

Lawnmower accessories

Five forward gears
give speeds ranging
from 1¼ to 4½ m.p.h.

Powerful engine unit
available in either 5hp or 8hp

Robust
steel construction

Single
heavy-duty blade

Wheeled steel deck
for large areas
of semi-formal
or rough grass.

In the same way that electric drills have a large number of accessories and attachments, from minilathes to drill sharpeners, there may be a need for a similar range of products to go with a petrol-driven lawnmower for the domestic market. Such products would enable the owner to carry out tasks with power tools that previously were laborious and difficult or required specialist tools that are expensive to buy and inconvenient to hire.

Specification Take any commercially available petrol-driven lawnmower (of any type provided that it is aimed at the domestic market) and imagine that you have been given the task of investigating the feasibility of producing a range of accessories to go with it, using the mower motor as the primary power source. Produce a list of possible accessories and a set of initial drawings of the attachment part on the mower and the mating connections on the various accessories.

Limitations on the design

Power Depending on the accessories that you think up, there will be some variations in the speed and torque requirements of the drive. Therefore, if your mower has a gearbox there is a decision to be made as to where to take the power off so that the requirement for extra gearboxes is minimised.

Cost If consumers are to buy the accessories they must be manufactured for less cost than an equivalent power tool with integral power source. Because of the reduced load-carrying requirements of the accessory it may be possible to make the accessory from a cheaper material. Plastics, for example, are almost universally used on electric drills where once there was much metal.

Environment The average lawnmower is frequently ill-maintained; having mown damp grass it is left in a damp garage or garden shed, unwashed, unoiled, and covered in sap and grass cuttings. Corrosion is a major factor in the design of garden equipment. Consumers do not like to see their expensive purchases rusting away after a few months of normal use, so efforts must be made to reduce the onset of rust and corrosion on the machines.

Corrosion is not the only problem. It must be remembered that not only do consumers not oil their machines, they do not possess spanners with which to make and break the connection between motor and accessory. Furthermore, these accessories are liable to be dropped, kicked, and otherwise maltreated (like vending machines, the more frequently they break down the less their expected useful life due to kick-induced damage!)

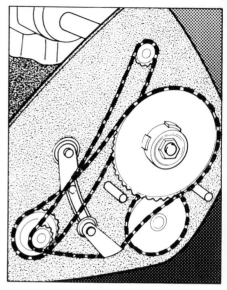

1 *Chain drive on a cylinder mower*

Safety Because lawnmowers are designed to cut a large volume of grass in as short a time as possible, there is a safety hazard to be overcome. It is the responsibility of the designer to ensure that his or her equipment is safe, even if not used in the way that he or she envisaged. In some countries the designer has a legal, as well as moral, responsibility to ensure safety. An injury caused by a piece of equipment can lead to the designer being sued personally by the injured party. Safety, then, is a primary consideration in all equipment, but particularly where the public and rotating machinery or cutting blades may come into contact with each other!

Accessories One of the first problems is to decide what type of add-on is suitable for a lawnmower. Lawnmowers are not as easily portable as electric drills, for example, so any accessories that are bolted directly on to the body of the mower should be semi-static. Were there to be a light umbilical cord to transmit power from the mower this difficulty could be alleviated.

Two avenues that might be explored are: putting one or more access points at certain

3 *A lawn rake*

places in the drive train; and, with self-propelled mowers, replacing the cutting element with modules that require a relatively large area to be covered fairly quickly, such as a lawn rake or aerator. It must be remembered that a mower produces only rotary motion; if some other form of motion or energy is required some intermediate form of power transfer might

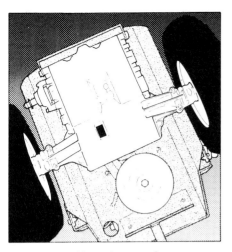

2 *You may be able to adapt a gearbox*

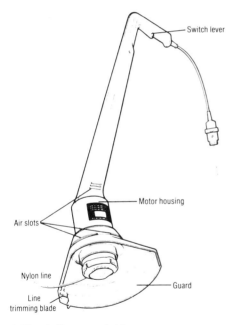

Switch lever

Motor housing

Air slots

Nylon line

Guard

Line
trimming blade

4 *Electrically powered strimmer*

be suitable. It may be necessary to add some extra hand controls beyond the throttle available on most mowers. If these are required some thought should be given to implementation.

Implementation The function of the accessory will, to a large degree, govern the type of attachment and amount of power

required. The power of mowers varies from about 400W for the electric ones up to about 6kW for some of the larger petrol-driven machines. This, of course, limits the power available for any tasks that the motor is expected to perform.

On the safety side, some of the cylinder mowers have an automatic clutch. It is in fact centrifugal, that is to say, as the motor speeds up, a set of clutch plates spin apart and take up on a drum connected to the output shaft, which causes the blades to start to rotate. It may be possible to adapt such a clutch so that it would disable the blades while allowing useful power take-up from the motor.

The self-propelled mowers already have some sort of power take-off to the drive wheels or roller. It may be possible to use this at very little expense and so improve the competitiveness of the system. If an accessory, such as a hedge-trimmer, needs to be portable, there are commercial torque transmitters available. They consist of a long, flexible, cylindrical member that is coupled to the output shaft of the motor and twists about its own axis so transmitting the torque. The whole thing is contained in a robust sheath. These can be heavy, expensive, and inefficient, particularly when power needs to be transmitted over more than a couple of metres.

Another way of transmitting rotary motion is in a manner similar to the old dentists' drills. A number of long rigid arms with pulleys at either end and cord round each pair of pulleys is linked together to form a long, flexible drive chain. The difficulty here is the limited amount of power available. This can be overcome to some extent by using chain drives or toothed belts, but this would appear to be a bit clumsy and not as safe as some other systems.

Teaching notes

As suggested in the body of this exercise, this idea arose from the multiplicity of tools and attachments available for the domestic electric drill. In general one would expect the accessories for a mower to fall into three basic categories: firstly, the semi-static ones – a post-hole driller or a water-pump; secondly, the power conversion units – a pneumatic or hydraulic compressor or electricity generator from which other units may be powered; and finally, the items that need to be portable – hedge-trimmers and so on.

This last category is particularly difficult to implement because of the problem of directly transmitting torque. As mentioned above, commercial units are available, but they can be very bulky compared with the equivalent power transmission of an electric cable. This weight would reduce the maximum reach from the primary torque generator, so reducing the flexibility of the system as a whole. Also, the torque transmitters tend to become inefficient over long lengths because of the friction generated between the flexible inner member and the outer sheathing. It appears, therefore, that accessories can be remote from the petrol engine only if a power converter is attached to the mower and the remote devices have an umbilical cord to carry energy (in the form of compressed air, electricity, or other means) from the motor.

If a power converter is used, for example a generator, then the output would ideally conform to the relevant standard ie 240V, 50Hz, or in the case of compressed air, 7 bar (100psig). If a student produces a generator and a standard 3-pin socket as his accessory system, it would be instructive to ask how he would cope with the voltage and frequency variations that would occur with heavy or fluctuating loads. If this problem is overcome, then it is surely the ideal solution, allowing a huge range of accessories to be used both in the normal domestic environment and when remote from mains electricity. The constraints of cost, power, and environment still apply of course, but it is conceivable that such a generator linked to a mower motor (detached from the mower, perhaps, to reduce bulk and weight) would fill a need in the camping market where there is a requirement for a few hundred watts of power for lights or radio receivers.

There are three types of mower: the air-cushion; the cylinder; and the rotary, that is to say vertical-axis and mechanically supported. Most of the hover mowers are mains powered, although there is at least one that is petrol driven. The cylinder mowers are mostly petrol driven, although one or two are electric. The rotary mowers are now almost all heavy-duty and petrol powered; the light-duty ones having been pushed out by the hover mowers.

Electric mowers are rated at 500 to 1000W, but motor mowers may be anything from 2 to 8hp ($1\frac{1}{2}$ to 6kW). This difference is due to the fact that most motor mowers have to propel themselves as well as cut the grass and also that small internal combustion engines are not very efficient. Most of the larger cylinder mowers have a clutch (called automatic by the salesmen, although it is centrifugal), but only on the very biggest tractor-type rotary mowers can a gearbox be afforded.

The hover mowers have plastic bodies, to reduce weight and therefore the power required to float, and this is the only form of structure on the mower apart from the handle. The cylinder mowers have a fairly solid chassis and thin steel skins to cover up the moving parts and other dangerous areas. The chassis would be suitable for mounting any extra equipment, but the panels may need to be redesigned. The most suitable, perhaps, are of the heavy-duty rotary type that have a low-precision aluminium casting to enclose the blades. This is quite rigid enough to support a good deal of extra weight without affecting performance of the machine in any way.

Wire strainer

When building a simple post-and-wire fence, one of the last jobs to be done is to pull tight the wires which run the length of the fence. If this is not done a slight local load on the wire, such as a small boy climbing over, will move all the slack to one place and the fence will not be very effective. Hence a device is required which can apply a load to the wire, pull in any slack, and hold the wire while it is secured.

Requirement The device must be capable of accepting single or multi-strand wire, plain or barbed. It should be able to be anchored to a fence post or another stretch of wire. It must be light and compact enabling one person to carry it across rough ground and use it without assistance. It must be rugged enough to withstand an agricultural environment with minimum maintenance and should be able to apply load in excess of 1kN.

Specification The agricultural environment is perhaps one of the toughest. Tools may be covered in mud, left out in the rain, and expected to work with little or no maintenance. Farm workers are used to handling many forms of machinery and

tools but do not expect to need special instructions, so the device must be obvious in its operation, perhaps after a brief demonstration.

When in use the device must not put kinks in the wire since these are difficult to remove. It must be operable at any reasonable height without excessive effort. All forces applied should act, where possible, along the line of the wire since any lateral forces would strain the fasteners on the posts.

Environment and use

Wire strainers do of course exist in various forms. One of these is illustrated on this page. The pull is applied via a lever to a long chain. The other end of the chain carries a device which can be used to grip a wire by a wedge action or will slip on to the chain to form a loop round a post (illustrated).

The lever end also carries a wedge-action wire grip and this would normally grip the wire being strained. The lever carries two spring-loaded claws which alternately grip the links of the chain as the lever moves back and forth. When the desired tension has been reached, the next claw to be engaged locks the chain until the wire has been secured.

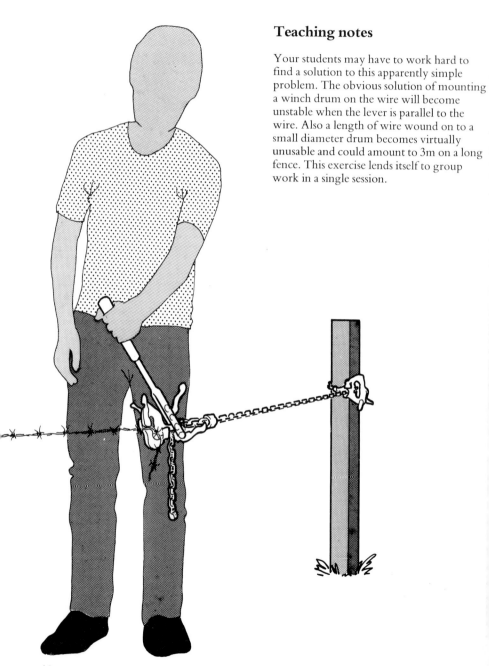

Teaching notes

Your students may have to work hard to find a solution to this apparently simple problem. The obvious solution of mounting a winch drum on the wire will become unstable when the lever is parallel to the wire. Also a length of wire wound on to a small diameter drum becomes virtually unusable and could amount to 3m on a long fence. This exercise lends itself to group work in a single session.

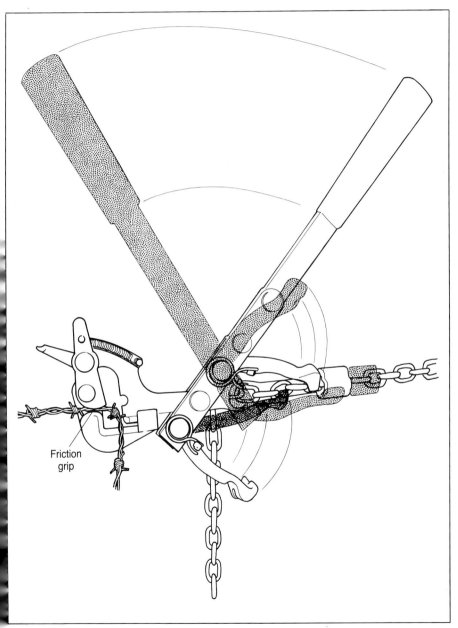

Friction
grip

The mechanism is simple but can generate great tension with small effort

Coffee-time challenges

These design exercises are all based on real industrial problems and all have real industrial solutions. They are called coffee-time challenges since they first appeared under that title in *Eureka* magazine. The ones selected seemed to the compiler particularly attractive in simplicity or economy – after he had seen the solutions! The problems, or quizzes as they are called, are given first and are followed by some possible solutions.

Quiz 1

Hacksaw blades, in the hands of professional and amateur alike, have a tendency to snap. New blades are not cheap and in any case the shops are usually shut when you need them most. The padsaw handle, which could save the day, in fact swallows all the remaining blade, leaving none to work with.

The problem Design a simple gadget to make the best use of a broken hacksaw blade; ideally, enabling the job to be finished.

Plodders will soften the blade, drill a new hole near the break, reharden and try to reuse it. But probably the hacksaw frame will not adjust to a blade this short! Besides, heat treatment over the picnic stove really is no fun.

Quiz 2

When samples of highly radioactive plutonium are moved by public transport in the United States, the 15in-diameter by 14in-high stainless-steel shipping containers used must be capable of surviving a stringent sequence of tests which simulate the worst possible air crash followed by a fire:
Impact: The container is hurled against a concrete target at a speed in excess of 288mph.
Crush: A 350-ton load is applied through a 2in-wide steel beam.
Puncture: A 500lb steel spike is dropped on to the container from a height of 10ft.
Slash: A 100lb steel angle beam is dropped

end-first on to the container from a height of 150ft.
Fire: The container is placed in a jet-fuel fire at temperatures above 1,850°F for one hour.
Immersion: The container is submerged in 3ft of water for at least eight hours.

The 15g of solid plutonium is located in the centre of the container in a cricket-ball-sized sphere made of high-strength iron-based alloy. The two halves of this innermost protective container are hermetically sealed and held together by 20 closely spaced bolts.

The problem Select the main material capable of containing the sphere centrally in the drum and able to withstand the sequence of tests. Note that thermal survival was the most difficult design challenge.

Quiz 3

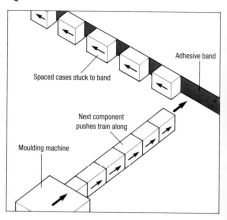

When handling large numbers of smallish components, cases for electronics components, for example, it is often convenient to fix them to a band for ease of handling during assembly operations. This is called 'bandoleering'.

One manufacturer needed to place rectangular plastics cases on to an adhesive band, similar to masking tape, with precise spacing for subsequent automatic assembly. The cases were delivered from the moulding machine at irregular intervals with each one pushing the component train along by a single unit.

The problem From the initial delivery train, design a means of spacing the cases accurately along the adhesive band.

Quiz 4

Measuring the amount of oil flowing through a plain bearing assembly is a useful way of monitoring the bearing's condition. With a constant pressure supply, the volume of oil flowing through the bearing will increase as it wears. When using this technique on large, expensive plant, such as power-station electrical generators or generators linked to marine turbines, a control orifice is used in the oil supply system to act as a reference datum.

The reference orifice needs a constant-volume flow of oil throughout the operating temperature range and, because of the oil's temperature/viscosity relationship, some means of control is required.

The problem Design a simple controller which will automatically allow a constant volume of oil to flow through it regardless of the temperature of the oil. The supply pressure is constant, but remember that an oil's temperature/viscosity curve is not linear.

Quiz 5

Before onions can be pickled they have to be topped, tailed, and peeled. The onions vary enormously in size, shape, and condition; and wastage, due to overcutting or overpeeling, is a real problem. Traditionally this operation has been performed by hand.

The problem Design an automatic machine

for topping, tailing, and peeling onions which can accommodate large variations in onion size, yet still remove the optimum amount of skin.

Quiz 6

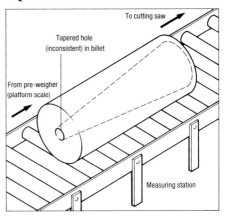

The diagram shows a 6in-diameter, semi-precious metal billet with an inconsistent, tapered hole along its centre line. Equal weight sections have to be cut from the billet for extruding into equal lengths of tube.

However the existing method of pre-weighing the billets and estimating the cut-off positions making allowances for the taper resulted in sections with either too little material, which were scrapped, or too much material, which caused unnecessary wastage.

The problem Design a system to weigh accurately each section before it is cut from the billet.

Quiz 7

The motor stator from a domestic central heating pump consists of a central steel tube around which is coiled a steel strip to form the necessary magnetic laminations. This subassembly is then pressed into a mild-steel cup to retain the coil tightly. The whole assembly is then gas welded together before milling 16 slots across the component.

Milling the slots produces burrs which have to be removed at great expense.

The problem Value engineer the component to simplify and reduce costs. Steel strip has to be used in the main body of the stator for functional reasons.

Quiz 8

Small, single-cylinder petrol engines have high inertia at the output shaft due to the inertia of the flywheel/crankshaft mass. In a garden lawnmower the cutting blades experience variable loads, remember that long wet grass, and may even jam at times when stones, twigs, and suchlike are encountered. To avoid damage to the engine and transmission, a shock absorbing element is often incorporated into the drive line.

The problem Design a simple shock absorber for use in this application. It should be easy to assemble and cheap to produce. It must transmit $1\frac{1}{2}$hp at 3000 rev/min and should be capable of accepting 2° of shaft misalignment in any direction due to production tolerances.

Quiz 9

Research in the food industry includes testing jellies. Investigations, for example, must establish the effect on jelly strength of factors such as concentration, pH, storage conditions, cooking, etc.

The problem Design a jelly tester. The machine should make the sticky-finger approach obsolete and yield quantitative, reproducible comparative results. First you must decide what it is that must be tested.

Quiz 10

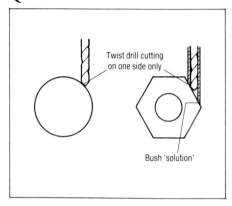

Twist drill cutting on one side only

Bush 'solution'

An acceptable way of locking nuts, round sections, and so on, on to a shaft is to wire or pin them in place, which often means drilling into an angled surface. In these circumstances the drill is difficult to start as it is cutting on its side and at one point only.

Even drill jig bushes shaped to fit the round or hexagon (as shown) offer no support for the point of the drill when starting. And even when jigs are used, breakages can occur when the drill is cutting out of the sloping surface because, again, it is only cutting on one edge.

The problem Design a means of supporting a conventional twist drill so that it will both enter and exit from the sloping faces of rounds or hexagons, without risk of deflection or breakage.

Quiz 11

Conveyor belts used to carry wet, sticky, corrosive or other difficult substances need to be kept clean if damage to the mechanism, belting, or transported materials is to be avoided. In the food industry, for example, any build-up could quickly lead to contaminated processing lines while, in other sectors, the conveyed material may cause extensive wear to the system.

The problem Design an efficient low-wear system for cleaning the return side of a conveyor belt. Conventional solutions, such as scrapers or brushes, both static and rotary, are ruled out because they cause accelerated belt wear, require excessive maintenance, and may even need separate motors and drives.

Quiz 12

The ability of fibre-optic cables to carry large volumes of data is now taken for granted. The basic fibre is not expensive: it is, however, fragile and considerable care is required when long cable runs are pulled through conduits and around buildings. Of course the cable can be reinforced and armoured, but this raises costs unjustifiably as, once installed, the armour is seldom required.

The problem Design a device that will grip the smooth outer surface of a fibre-optic cable tightly enough to enable it to be pulled through a building. It must work on a variety of cable diameters, must not apply point loads to the cable, and be capable of negotiating bends in narrow conduits. To add to your device's appeal, how about trying to design it so that it can be used when both cable ends are inaccessible, such as when installed cables require retensioning?

Quiz 13

Domestic robots, as promised by so many science fiction writers, are just beginning to appear. They cannot do much yet: mainly they wander about uttering simple phrases and the prospect of a robot which could wash the dishes or make the beds still seems a good way off. Opportunities abound for the inventive designer: even the simple matter of making a robot stand upright is far from closed.

If the domestic robot is going to be roughly human-shaped, it will be a long object standing upright on a small footprint. Four-wheel drive, caterpillar tracks, and the like must have large footprints to achieve stability, but human-like feet are hard to manage.

The problem Can you design a mechanism with a small footprint, which lets the robot rove without falling over?

Quiz 14

In the United States the toilet accounts for 45% of household water consumption. In hotels as much as 90% of water consumption originates there. With water costs rising at more than 25% each year there is a huge incentive to save water.

The problem Design a water flushing unit for a standard toilet bowl that will significantly reduce the bowl's water consumption without losing extraction efficiency. For the non-expert, most American – and half Britain's – toilets work by creating a siphon in the bowl's trapway. The flush provides sufficient flow to maintain the siphon for a short period and then refill the bowl.

Quiz 15

Not all automatic tea-making equipment for domestic use correctly fulfils the long-standing tea-making tradition. For early or late risers to appreciate fully their morning 'cuppa', not only is a state of consciousness necessary but the pot must be warmed. It is the fact that the tea is not transferred to a warm pot that gives tea-making equipment a bad name.

The problem Can you design an automatic machine, suitable for the domestic consumer market, that boils water, introduces to it tea or coffee (in leaf, bag or ground form), prevents further boiling while the liquid 'brews', and, finally, stores the drink in a warmed pot until required?

Quiz 16

Circular saws are used for pattern-making, trimming, cutting up stock material, and so on, often on equipment which has a guarded blade protruding through a worktable. How much better it would be to be able to use an unguarded tool from above that cuts the material but not the surface it is resting on, or the person doing the work.

The problem Design a power saw that will cut plastics, including GRP, laminates, and wood but will not, however, so much as scratch the operator should he accidentally come into contact with the blade. In addition it must be able to cut hard materials without damaging any soft subsurface (such as foam) and should not generate clouds of airborne dust.

Solution 1

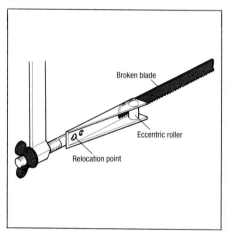

A sound idea comes from Michael Bilney of Maidstone, who has invented and patented the Bladesaver illustrated here. It is intended to allow the maximum possible use of the remaining length of blade.

The idea could also be useful in the machine shop. Too much money can be lost when a power saw is out of use for days, through a lack of spares. This might even be a better way to fit the blades in the first place.

Solution 2

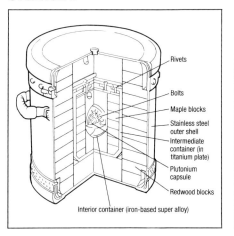

Rivets

Bolts

Maple blocks

Stainless steel outer shell

Intermediate container (in titanium plate)

Plutonium capsule

Redwood blocks

Interior container (iron-based super alloy)

Solution 3

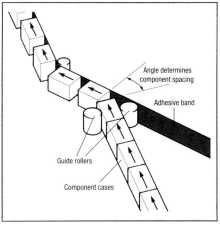

Angle determines component spacing

Adhesive band

Guide rollers

Component cases

Project engineer John Andersen's solution was to make extensive use of wood because of its good energy-absorbing characteristics.

In early designs only redwood, which has the highest specific energy absorption rate of any shock mitigating material, was tried but, under impact tests, the interior sample-holding ball was able to tear through this wood. Maple, which has greater compression strength, was substituted in the inner section of the final design.

An additional advantage of wood is that when it burns it converts to a carbon char which has insulating properties similar to the heat shield of a space vehicle.

Sandwiched between the two wood layers is a $\frac{1}{4}$in thick titanium container which helps to spread impact loads to the grain-orientated wood and dissipate heat generated by the payload. Initially this inner container was made of aluminium, but it did not survive the thermal tests.

Engineering designers Rhoden Partners Ltd of Acton, London, solved the problem by introducing a change of direction in the train of cases. By causing them to turn as they feed on to the adhesive band, a gap is opened up the width of which is dependent on the angle of swing (see illustration).

The ejecting components not only push the train along, but also drive the adhesive band through component contact. In this design irregular delivery does not affect spacing as would the more usual differential speed conveyor system.

Solution 4

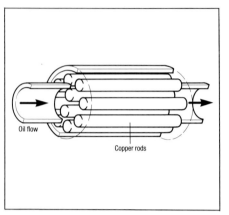

Oil flow

Copper rods

Designers at Ampy Automation arrived at the following solution only after much mathematical analysis.

They fitted a set of copper rods into a tubular manifold and made the oil flow through the interstices. As the oil heats up, the copper rods expand. The consequent constriction in oil flow, a result of the reduced area, exactly matches the increase in flow brought about by reduced viscosity.

It was found that over a working temperature range of 40–120°F the flow rate remained constant to within ± 1%.

To achieve such accuracy the frictional changes in the manifold have to be taken into account along with the area reduction.

Solution 5

A machine meeting this specification has been designed by Peter and Robin Jones of R Jones (Engineers) Ltd of London. The onions are held between, and conveyed by, two belts. Self-adjusting guides sense the size of the onion and adjust the cutting position of rotary topping-and-tailing knives.

Further travel causes the onion to pass into two stationary hinged knives, one on each side of the conveyor. As the onion is pushed through these knives, they draw across the onion face making a cut approximately one skin-layer deep. This cut serves as an inlet for high-pressure water jets which peel and wash away the outer skin.

The water-jet skinning is independent of onion size or shape and provides a gentle method of peeling.

The machine is handling 1000kg of onions a day. This is a dramatic improvement compared with the daily output of 25kg from each operator when the process was done by hand.

Solution 6

The billet is supported from an overhead hoist and lowered into a full tank of water. By incorporating a strain-gauge load-cell into the hawser, it is possible to monitor the apparent change in weight of the billet as it is immersed in the water, which overspills.

Using Archimedes' principle the weight of the immersed portion is monitored by a weighing instrument which triggers a marking gun at water level. Accurate instrumentation is necessary as the specific gravity of the material being weighed is many times that of water. In addition, the weight change due to the water's displacement is small, but it is not difficult to achieve an accuracy of ± 0.02% using load cells.

This method of weighing is obviously directly related to the cut section's volume, and minor variations in the material's specific gravity and in marking and cutting errors have undesirable effects. Nevertheless, in this particular application the scrap rate was reduced by around half.

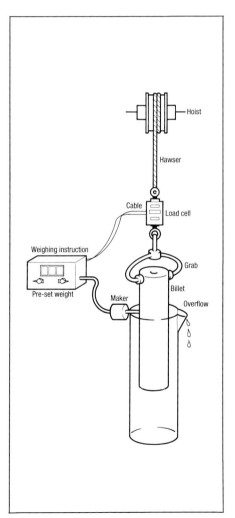

Solution 7

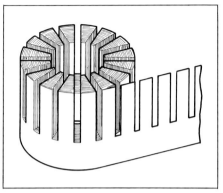

Sealed Motor Construction Company Ltd of Bridgwater reduced from 15 to just two the number of operations needed to produce the component illustrated.

The coil is punched and rolled in one operation with the component's increasing outer diameter being continually sensed by a special system. This system triggers the punch and ensures that the stator slots are produced at precisely the correct pitch irrespective of variations in material thickness to form perfect, straight-sided slots across the component after winding.

Spot welds on the inner and outer strips hold the coil together and the tube and cup, previously needed to reinforce the strip during milling, have been eliminated.

Solution 8

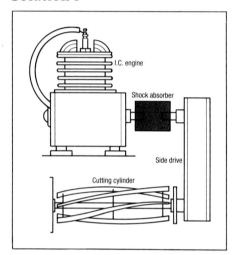

There are many ways of using a rubber cushion, chosen because of its high deflection and good damping qualities, but the answer we like is from Qualcast Ltd. In one of their mowers, designers used a piece of reinforced rubber hose, similar to car-heater hose, to connect the two shafts together. The hose fitted snugly over each shaft end and was held in place by two jubilee clips, one on each shaft.

The frictional contact developed was sufficient to transmit the drive and any massive shock overload allowed the hose to slip. Normal shocks and misalignment were absorbed by deflection in the hose. This answer worked well, it was compact and, of course, cheap on spares.

Solution 9

And the answer is – a bucket and spade! On the FIRA Jelly Tester the spade is inserted into the set jelly to a specific depth. It is then required to measure the torque needed to rotate the spade through a fixed angle. The torque is exerted through a pulley system by a water bucket filled from a slow source (100ml/min). The instrument, made by a Croydon company, meets the requirements of BS647 and BS757 which deal with methods of testing jelly strength for comparative purposes. A sticky finger probably would not.

Solution 10

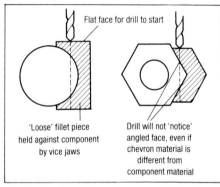

Using a blanked-out fillet piece, either chevron or crescent shaped, the job becomes a simple matter of drilling into a flat surface. Nor is there a pilot hole in the fillet piece, so drilling commences on the level, flat surface in the ideal conventional manner. Provided that the fillet is pressed tightly against the component, the material change at the join line is of no importance and the drill passes cleanly into the component. This also applies as the drill breaks out.

Apart from speed of operation, claimed by one manufacturer to be half that of any other method, deburring is either eliminated or considerably reduced. One manufacturer,

Valeport (Services) Ltd, markets a special jig and chevron manufacturing service for sizes up to 2in AF. Using this tool the chevrons, held in a magazine, are dispensed and held automatically under hand-lever control.

Solution 11

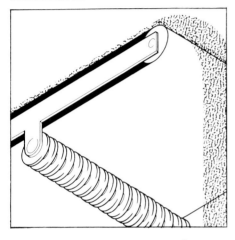

Interroll spiral return rollers are made up from separate, moulded rubber discs mounted on a plain shaft. These are in contact with, and are driven by, the moving conveyor belt itself; a process which provides a constant and steady cleaning action through the flexing of the rubber discs as they turn. In addition the rollers not only support the belt during its return cycle, but also help to maintain belt alignment. The developers tell us that standard rollers can be used in ambient temperatures from -20 to $+80^0$C while 'specials' can be supplied in oil- and flame-resistant materials.

Solution 12

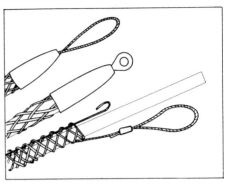

Remember the Chinese handcuffs in Christmas crackers? Insert your fingers in the woven tube and no amount of pulling will release them.

Designers at Harvey Hubbell used this principle in the design of fibre-optic-cable grippers. An uncomplicated woven wire mesh slips easily over the cable end. The harder it is pulled, the tighter it grips, but always over a large surface area so that it never bites into the cable. A reinforced metal nose helps the cable puller to negotiate bends and the gripper itself reinforces the vulnerable front end of the cable during installation.

For retensioning applications, a flat mesh is wrapped round the cable and closed by threading a metal pin through the mesh's loops.

Solution 13

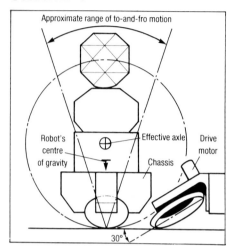

Approximate range of to-and-fro motion

Robot's centre of gravity

Effective axle

Drive motor

Chassis

30^0

The outstanding solution invented and patented by Douglas Jones of the California-based Androbot Inc amounts to no less than reinventing the wheel. He has developed a two-wheel system. His arrangement has a 3ft-high robot sitting stably upright on a pair of 10in wheels and it will even bob back to the upright when knocked as much as 15^0 away from the vertical.

The secret is that the wheels are tilted over to make an angle of approximately 30^0 with the floor. Viewed from the side, the wheels now look like shallow ellipses with quite a large effective radius at the point of contact with the floor. That means that the dynamics are similar to using large wheels, with a diameter nearly equal to the robot's height.

All that is now needed is to arrange for the robot's centre of gravity to be below the imaginary axle of the large wheels, and the device is stable, and can recover from tilting, within limits. In fact, it tends to rock back and forth a little as it moves, and customers, we are told, seem to find that amusing.

Solution 14

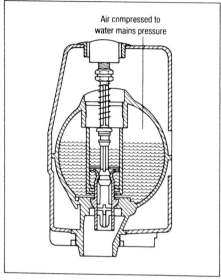

Air compressed to water mains pressure

Control International in the United States discovered that almost half the flush water volume was used to start the siphonic action in the bowl. This was because the gravity-fed water pressure could not produce high enough flow rates. So they designed a self-pressurising water canister that discharges the flush water under pressure.

Because of the substantially higher flow rate the siphonic action starts immediately and water savings of between 35% and 75% per flush are possible. The Cashsaver DX-1400 unit uses the principle of air compression. Supply water enters a sealed vessel until the trapped air becomes compressed to the same pressure as the mains supply. When the flush valve is triggered, the water is released at mains pressure: about 40 to 60psi. Much careful design has gone into the unit since it has to work reliably in a hostile environment under pressure for many years.

Solution 15

Designers at Russell Hobbs literally used a boat to solve the problem.

Their tea-making machine uses a ballasted moulded plastics container, with a perforated lid, to hold either tea leaves or coffee grounds. This floats with its dry cargo on the water in an electrically heated water jug.

When the water boils its surface becomes so agitated that the boat ships water and very quickly sinks. The ballast used is a magnet and it is this, once the submerged vessel approaches the bottom of the jug, that triggers a microswitch which turns the heating element off.

Having introduced the dry ingredients to freshly boiled water, the two constituents can be left to brew for as long as desired; the heating jug now doubles as a teapot or coffee-pot.

The design also incorporates certain safety aspects so that an empty jug containing a boat cannot be switched on: the microswitch is held in the off position by the magnet. And it is impossible to boil the jug dry, provided the boat is not omitted.

Solution 16

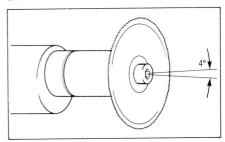

Anyone who has ever had a plaster cast removed in hospital may have come across Desoutter's answer to this problem, possibly without realising it. The cast cutters used look exactly like circular saws but are, in fact, quite different. Instead of rotating, the blade oscillates at high speed through an arc of 4° allowing safe, controlled cutting.

Industrial versions are now also available, including a pneumatic type that can be used in flame-sensitive areas with complete safety.

This version oscillates through an arc of 4° at 14,000 cycles/min and accepts a variety of different blade shapes, including circular, for high-accuracy profiling, plunge cutting, and so on. Also of course, because the blade is not rotating at high speed, very little airborne dust is generated. All dust is simply removed using an extraction hood connected to a vacuum cleaner.

Seeking a market

The design process begins with the identification of a market need which can be satisfied. Usually these needs are identified by an analysis, often informal, of customers' requirements coupled with a knowledge of the capacity to achieve functional performance. This is the 'market pull' approach which leads on to a broad based conceptual phase exploring ways of fulfilling the need. But the design process frequently begins from a completely different starting point. A new piece of technical information may arise from a piece of research; or a technique well established in another field may be applied to a different problem; or even a little known 'scientific curiosity' may suddenly assume a useful aspect (electricity was once in this category); and suddenly there is a potential solution available for an unknown market requirement. This is known as 'technology push'.

Design exercises are normally formulated in market pull terms since these are more straightforward and possible routes to a solution may already be known or easily identified. Two exercises are given here based on technology push using two scientific curiosities which have been known for some time. The end results are completely unknown at the beginning of the exercise and thus present a greater challenge to both students and staff, although, for the latter, formal assessment may be impossible.

Ionised flames

Figure **1** shows a simple rig which demonstrates an unusual effect. A gas blowtorch heats a thin non-metallic sheet of material, behind which is a metal plate which does not necessarily touch the sheet. Electrical potentials may be applied to both the torch head and the plate which must be otherwise electrically isolated. Under normal conditions, with no applied charges,

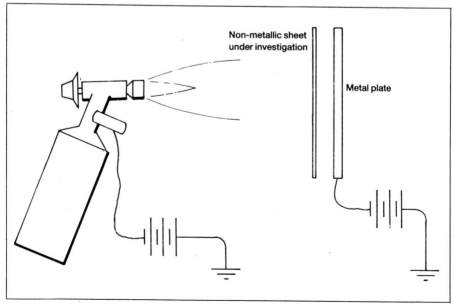

1 *Laboratory demonstration of ionised flame*

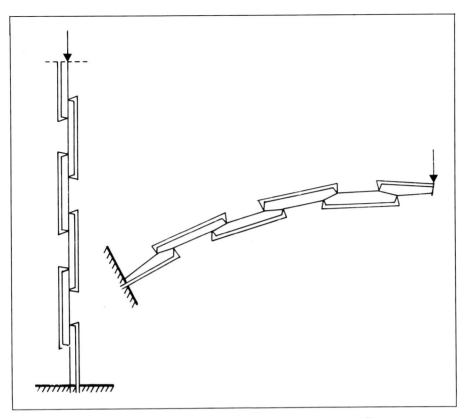

2 *Pre-tensioned column*

3 *Demonstrating lateral stiffness*

the heat from the flame will soon char or otherwise affect the sheet material.

If voltages of the same sign are applied to both the torch head and the plate, the heating effect is reduced or even eliminated depending on the exact configuration, dimensions and voltages used. (Care should be taken to avoid dangers from both the flame and the charged parts.) If the voltages are of opposite signs the heating effect is increased. These effects are due to the ionisation of the gas in the flame, which is then repelled or attracted by the plate. You may carry out your own experimental investigation to explore the effects of varying the distances from the torch to the sheet and the sheet to the film, varying the applied voltages, changing the shape of the plate and the thickness and material of the sheet.

Many processes involve gas flames and the ability to change the heating effects by electrical means opens up some interesting

possibilities for useful applications. Just finding a potential application is not enough. You must then identify some potential customers and design a product to meet their needs. The conceptual part of the design process here effectively precedes the complete identification of the market but the rest of the process is the same as for market-led exercises.

Stretched columns

A cable or bar in tension has a strength limit determined entirely by its cross-sectional area and completely independent of its length. In compression the problems of buckling, requiring a consideration of section modulus and length, severely limit the strength often to only a small fraction of the tensile strength. Figure **2** shows an arrangement which compensates for some of the compressive deficiencies by using a series of short compressive struts to apply a tensile force to a cable in excess of the applied compressive load. Since the struts are short, and their compressive strength varies inversely with the length squared, they can be of a much smaller cross-section, and hence weight, than a single full length member to carry the same load. Of course there is a penalty. The structure has buckling modes which depend on the amount of overlap of adjacent struts and the relative stiffness of cables and struts. The assembly has a lateral stiffness as can be seen in figure **3** where it is shown acting as a cantilever. The deflection is exaggerated for a practical structure although a simple wire and fishing line model will show this order of movement. The analysis, particularly of the bending mode, quickly becomes non-linear and an experimental exploration may yield faster results than an analysis for a particular configuration.

This type of structure may prove lighter under certain conditions than the equivalent simple column of the same material. Applications, where a folding or easily erectable structure are needed, may be worth consideration. Trusses, where these assemblies are the compression members and simple cables are the tension members, could also be explored. All of this is unknown and the remarks about ionised flames apply here just as much. The only known application so far is the type of model shown in figure **4** occasionally found as a engineers' executive toy!

Teaching notes

This form of design exercise is inherently more difficult to assess since the results, if any, are completely unknown at the beginning. The two examples given here, although valid in themselves, are intended to show what might be done with those fascinating snippets of technical information which have yet to find a use. Many useful and important products or even entire industries have grown from such

investigations. Technology push is responsible for the present computer industry where progress is limited by finding a use and market for the new technology. Many less significant design projects in industry begin in this way and the exposure of students to the different approaches needed is an important part of their education.

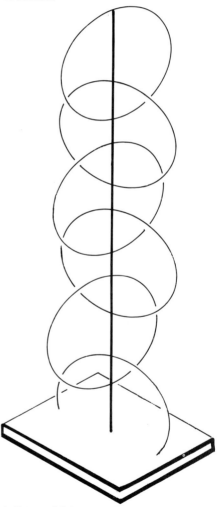

4 *Wire and fishing line executive toy*

Design Investigations

Sun awning support

Most shops, particularly those selling goods likely to deteriorate in strong sunlight, have some form of awning over their main windows to shade the items on display. This awning is usually a roll of canvas material set in a slot above the window. It is unfurled and supported on a folding structure, which is cantilevered from the wall and folds away to be unobtrusive when not in use. This design investigation looks at those folding structures. There are a number of types, four of which are shown here. Each comes in a number of variations depending on the awning size and any other special considerations.

Students should explore their local shopping areas to see how many different types of support they can find. These should be sketched or photographed and, if possible, measured and checked for their method of operation. The different types should then be catalogued and compared as to their basic design principles. There is scope for drawing and model building here. Students may like to try their hands at designing their own devices to do the same job. A design should support the awning without interfering with pedestrian access below it. One person should be able to open and close it quickly and easily without excessive effort.

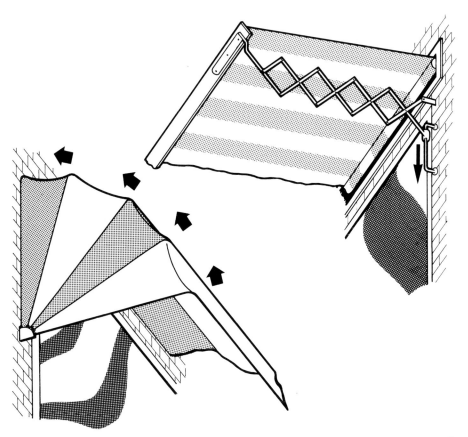

This work can serve as an introduction to other types of folding structure such as umbrellas, simple and folding; tents, ridge, frame, marquee, etc; folding garden chairs and tables; and inflatable structures such as boats and camping mattresses. Common mechanisms such as lazy tongs, concertinas and 'knee-over' locking stays should be noted and their working principles investigated. Space projects provide a rich source of folding structures to support solar arrays, radio antennae and stabilising booms.

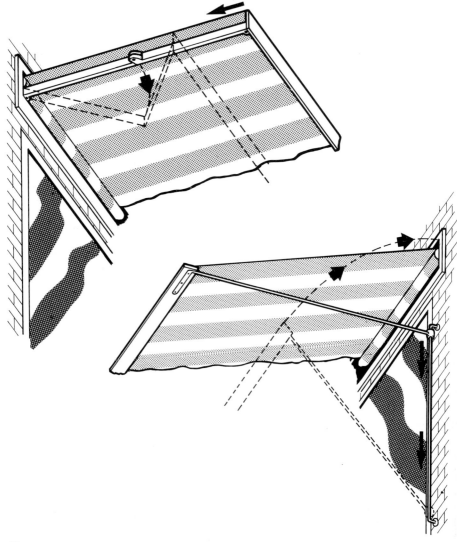

Communicating by light

The latest form of light industry is precisely that. The development of fibre optics, lasers, sensors, displays, and so on is revolutionising the communications industry. These new technologies will be significant in the communication of information over short, medium, and long distances. From liquid crystal displays to transatlantic telephone cables; from cable television to optically based computers; from photomultiplier tubes to lasers and masers, they are all products of the opto-electronics and electro-optics industries.

The distinction between the two lies in the roots of the technology. Sensors and displays have long been part of the electronics business, but only now are companies and countries developing new forms of electronic/optical transducers. These are usually called opto-electronics. Electro-optics, on the other hand, are the systems that are rooted in the optical sciences. Lasers and fibre-optical phenomena come into this category: essentially the areas where light is required to travel over more than a few metres.

The distinction is necessarily vague, as the two areas overlap considerably and will do so to a much greater extent when optical fibres are used to transmit information between countries and continents in the near future. The greater bandwidth available at optical frequencies compared with radio frequencies enables much more information to be carried on a given cross-section, opening up the possibility of video telephones and other futuristic developments.

Display technology

The story of the development of low-power displays is told in the development of pocket calculators.

One of the first mass-production calculators to be available had a five-digit light-emitting diode (LED) display. It drank power. It gave out radio-frequency interference (RFI). The segments were tiny.

A modern calculator bought for the same price has a nine-digit liquid-crystal display (LCD). The single small lithium battery will last for three to four years without replacement. The segments are large and legible in any light where one can write with a pencil. There is no RFI. Furthermore the display may contain dedicated symbols or words that describe the status of the calculator and switch on and off accordingly. Of course these are not the only types of display, though they are the most common in compact equipment. The most widely used in general equipment is the television screen or cathode-ray tube. This is very flexible, capable of high resolution, and very widely available; however, it is quite bulky and not cheap to produce. The bulk is steadily being reduced as more applications are developed. At least two manufacturers have already announced flat-screen television displays that rely on conventional display technology, but introduce a new

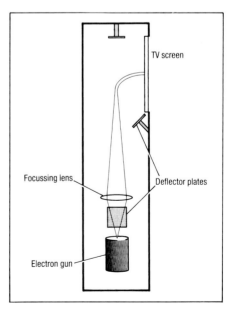

1 *Flat-screen television geometry*

geometrical configuration. The new geometry includes a display screen that is parallel with, rather than perpendicular to, the primary axis of the tube.

Graphics display tubes

Until there is a radical new development in high-resolution graphics displays the new screens will still use one of three main types of display technologies. These are all used in the computer-aided design and computer-graphics industries and may be classified under three headings: raster scan, vector scan, and direct-view storage.

The raster-scan tube is the most familiar, being based on the ordinary television tube. That is to say that any image is made up of a matrix of discrete dots, so that a diagonal

2 *Raster-scan video screen*

line is made up of a series of short horizontal and vertical lines. This reduces the clarity and resolution of the display. A television screen demonstrates this quite well, but monitors designed for use with computers generally have a higher resolution; the matrix is much finer and the electron beam is directed much more accurately at individual phosphor dots. Raster-scan displays contain a memory and each

memory element corresponds to a single picture element. Reductions in raster tube prices are often linked with reductions in the price of semiconductor memories. The greatest advantage of these tubes over others is their capacity to fill in large blocks of space in a variety of colours.

The vector-scan tube is like the raster tube in that the beam refreshes the image every few milliseconds, allowing the display to be updated in real time. The great advantages of the vector-refresh screen are the resolution and clarity and a capacity for drawing smooth curves. The cost is high, however, both financially and in the limited capacity for displaying colours. The vector-scan tube can display coloured lines, but has no capability of displaying blocks of illumination either in colour or monochrome. The greatest financial cost is derived from the requirement for high-speed analogue circuits to drive the electron beam around the screen in the shortest possible time. There is also a necessity for substantial local processing power to traverse the display data and to communicate this to the screen.

The direct-view storage screen comes from the same family as the vector-refresh display; such screens were first used in Tektronix storage oscilloscopes. The (monochrome) long-persistence phosphor screen is flooded with a diffuse beam of electrons, not sufficient in itself to illuminate any part of the screen. When a finely focused electron beam writes to the phosphor, the flood will sustain the image for many minutes. Although this system is fairly cheap, it does have a number of drawbacks, apart from the lack of colour information. Chief among these is a requirement that the whole display is redrawn every time a change is made to any part of the screen.

New developments

The race is now on to get away entirely from the electron-beam approach and produce a solid-state display based on liquid crystals or other technologies. Already IBM has announced a preliminary electrochromic display using viologens: substances that have previously only been used for simple seven-segment numerical displays. The goal is a high-contrast display that does not need to be refreshed frequently, but may be refreshed selectively at any time, that is to say, capable of a grey scale and using the minimum of power at voltages compatible with standard logic circuitry. As this goal is approached and the number of addressable picture elements (pixels) rises significantly above 500 × 500, high resolution graphics will be available in many products now limited to less-sophisticated display technologies.

The most obvious market for a cheap, high-resolution screen is the home computer field. Most home computers rely on UHF television screens which can only support 40 characters across a screen width. It is generally acknowledged that if a text-processing application is to be useful, the screen must be capable of displaying at least 80 characters. This requires a resolution of approximately 1000 × 1000 pixels.

Opto-electronic sensors

The very first displays, single leds, were products of the market for detectors. A simple optical sensor consists of an emitter capable of a fairly bright output and a receiver capable of distinguishing between background and signal radiation. Many of these sensor-emitter pairs work in the infra-red band, over very short ranges (mm). Infra-red radiation is produced by gallium-arsenide leds. Others, using lasers and laser detectors, have bounced signals off the moon. Normally, however, such sensors are used over distances up to a few metres. The vast majority of all such devices are used to detect any interruptions to the projected beam. They may be used for safety, security, or control, also for measurement, counting, or quality assurance. Such sensor pairs are very versatile and cheap, costing only a few

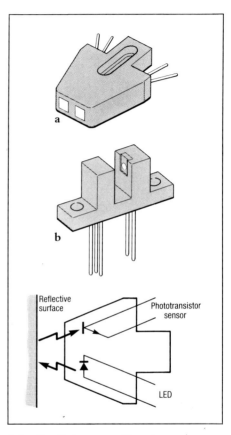

3 *Small optical sensor-emitter pairs are cheap and versatile. They come in many shapes and sizes:* **a** *and* **b** *are two of the most common*

pounds for each sensor-emitter pair.

During the development of digital electronics it became obvious that data was becoming corrupted by spikes on the lines that carried the data. The sensor-emitter pairs were put to use as isolators. Because the emitter could be switched very rapidly, at rates equivalent to rates of data transmission, streams of digital pulses were fed into a led which pulsed on and off accordingly. A spike during the off state was too fleeting to switch the led on; a spike

69

during the on state made no difference to the output of the led. The power supply to the detector is clean of all spikes and the detector picks up a stream of clean digital pulses. Opto-isolators are now fairly common chips, particularly in electrically noisy environments such as motor cars or machine tools. Solid-state relays designed to

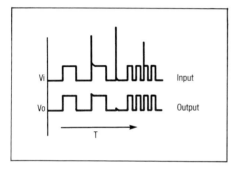

4 *An optical isolator removes unwanted interference*

switch mains voltages from logic signals are almost all optically isolated to reduce the possibility of reactive voltage spikes appearing on the signal lines.

Electro–optics

If electronics was the growth area of the 70s and 80s, electro–optics stands to be the growth area of the 80s and 90s. At present the field is more or less covered by fibre optics, lasers, some new forms of displays, and, depending on application, video tubes, image intensifiers, cameras, and so forth.

Lasers, after an initial development period in the 50s and 60s, had something of a reputation for being solutions looking for problems. Now lasers have military uses in range-finding and target spotting, but the difficulties of aiming a high-power laser over many kilometres are formidable. In the civilian environment many companies use lasers to cut patterns in sheet metal or to trim pipes. The advantages are great: a small heat-affected zone, almost burr-free cutting,

the capacity to cut very intricate patterns in very thin foil, flexibility, cleanliness, ease of use, and so on.

Because a laser can be used as a delicate cutting instrument, there are medical applications, particularly where work is being carried out in a very confined space, such as eye surgery. The beam cuts by heating and the heating effect automatically cauterises the tiny blood vessels as they are cut, so that very little blood emerges to obscure the view of the surgeon.

The chief properties of the laser may be used individually or in combination to solve different problems. The beam is parallel, coherent, and monochromatic; it also has a high-power flux density. Many range-finding applications make use of the coherence of the beam. The laser beam is split into two halves, one of which, the reference beam, is directed over a precisely known distance. The other half is directed at the target and the phase difference between the reflected half of the beam and the reference beam can be used to calculate the distance of the target. This phase differential is the basis of holography and laser gyroscopes. Laser gyroscopes use two beams, both of the same frequency, travelling in opposite directions around a triangular path created by mirrors. If the gyro is rotating about an axis perpendicular to its plane, one of the counter rotating beams will have further to travel than the other, this will cause a change in the interference pattern which can be detected by sensors around the vertices of the triangle.

Thus the laser gyro can give out signals to a computer and thence to a series of servomotors to correct any undesired spin. However, it will not act like a normal gyro and produce a Coriolis force perpendicular to both the axis of rotation and an applied force, neither is it quite so sensitive to small dust particles and the like in hostile environments.

Some of the more interesting applications require monochromatic light. Sodium light

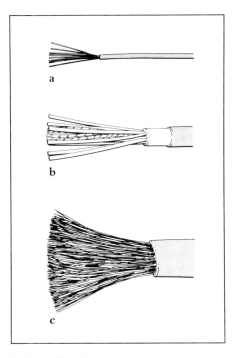

5 *Optical fibres (***a***) can carry much more information per unit area than copper cables (***b*** and ***c****)*

is almost monochromatic, in fact the two emission lines are barely 0.6nm apart, but a laser, whether carbon-dioxide, helium-neon, YAG or ruby generates what is, for all practical purposes, a single optical frequency. This is of great importance in the communications industry where the number of telephone channels (for example) that can be carried depends partly on the bandwidth of the information to be transmitted, but also on the number of carrier frequencies that can be sent simultaneously down a single conductor. If the frequency spread of the carrier can be restricted sufficiently, then the clarity of the transmitted signal, and also the number of individual signals that can be transmitted, increases enormously (see figure **5**).

Hinges

One of the simplest mechanical components is the pivot. A device which allows rotational movement in one plane and yet provides a load-carrying capability across the pivot. The familiar hinge is a particular form of pivot where the range of rotational movement is restricted to less than one revolution, and it usually has the capacity to carry loads along the pivot axis as well as across it. The object of this investigation is to look at different types of hinge mechanism, ignoring differences in fixing plates, decoration, and scale.

Pin hinges

As the name implies, this type of hinge contains a pin about which the other parts revolve. The pin acts both as a bearing surface and as a load-transfer member. The simplest form is the gate hinge where a

vertical pin, protruding from a bracket on the gate post, supports a metal strip with a loop or hole in it attached to the gate. This has the advantage that the gate can be lifted off if required. These hinges are used in pairs, but only one of them carries the weight of the gate (figure **1**).

A variation of this can be found on farm gates in some parts of the country, where the bottom hinge is replaced by the three-pin system shown. With this arrangement the gate pivots on a different outer pin for each direction of opening. Because these pins are not directly below the top hinge, which carries all the weight, the gate is lifted slightly as it opens. This makes the gate self-closing from both directions. The third central pin is added to give a small normal arc of movement and so avoid the problems of aligning the posts exactly.

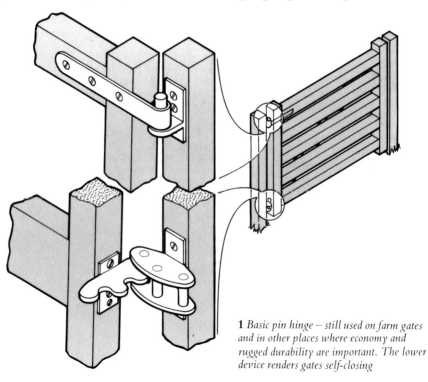

1 *Basic pin hinge – still used on farm gates and in other places where economy and rugged durability are important. The lower device renders gates self-closing*

Butt hinges

For doors the pin hinge has evolved and gained extra loops to become the familiar butt hinge (figure 2). This also has a gravity-closing variation, the rising butt hinge (figure 3). The rising action is useful for clearing thick carpets and the hinge is often fitted for that purpose. There are also falling butt hinges where the helix winds the opposite way making the door self-opening. When fitting these hinges care must be taken to allow room for the door to move vertically in its frame or the door will jam.

A double-acting gravity-closing hinge for lightweight doors, which also uses a helix is shown in figure 4. Here the bottom hinge carries the weight and has the helix mechanism while the top hinge simply carries horizontal loads on a pin which rises and falls with the door.

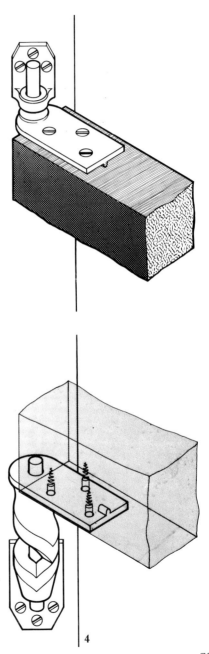

2 *Basic butt hinge – made in a wide range of sizes and load capacities*

3 *Rising butt hinge – uses a simple screw action to lift or lower the door, rendering it gravity-closing*

4 *Two-way variant on the rising butt theme, mainly used for decoration*

Multiple hinges

A double-acting hinge with only one pin
has a restricted range of movement because
the door hits the frame if it moves much
more than 90° from the closed position. For
some doors and for folding screens and in
other similar situations it is necessary to have
a 180° movement in either direction. This
can be obtained from the double hinge
shown in figure **5**. Here there are, in effect,
two sets of links working in opposite
directions. If only one middle leaf was used,
the hinge would simply flop open. With
two sets of interleaved links, only one
vertical line of pins can open at once. If
strong springs are employed to keep the
unused leaf in place, a self-closing version
can be devised (figure **6**). This is a much
stronger and more useful hinge than that in
figure **4**.

On furniture such as cupboards and
cabinets there are often two doors which
hinge on the same frame. Here a double-
door hinge (figure **7**) uses the same pin for
both doors.

5

5 *Double-pin hinge with an extra centre leaf
that is split to give correct movement*

74

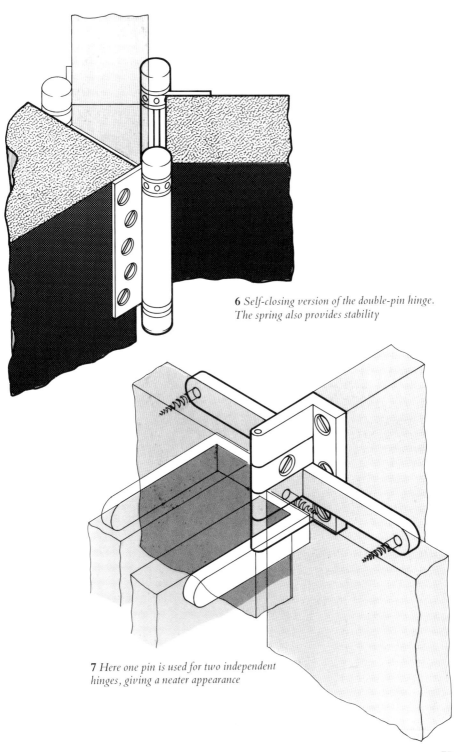

6 *Self-closing version of the double-pin hinge. The spring also provides stability*

7 *Here one pin is used for two independent hinges, giving a neater appearance*

Concealed hinges

Other furniture styles call for concealed
hinges. This presents a more difficult
mechanical problem since a single-pin hinge
has to have its pin protruding in order to
allow the door to open fully. Hence a more
complex mechanism is required to create an
apparent pivot without an actual pin. One
hinge using interleaved sliding members is
the SOSS invisible hinge which is actually
recessed into the edge of the door (figure 8).
This uses a small device with five pins
holding a set of interleaved plates together
to form a sliding mechanism with a very
compact form when closed.

Where it is acceptable to put the
mechanism on the inside of a cabinet it is
possible to devise a mechanism which will
allow the door to be flush with the frame
when closed and flush with the cabinet side
when open (figure 9). Another use for this
type of lazy-tong hinge is on table flaps
where a flat top is required. Note that these
hinges must be made for specific thicknesses
of door or flap.

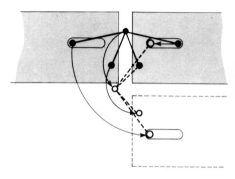

8 *The* SOSS *hinge folds neatly inside small holes
when closed, but forms a strong load connection
when open*

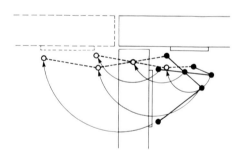

9 *This lazy-tong mechanism positions the door,
or table-leaf, in two specific positions*

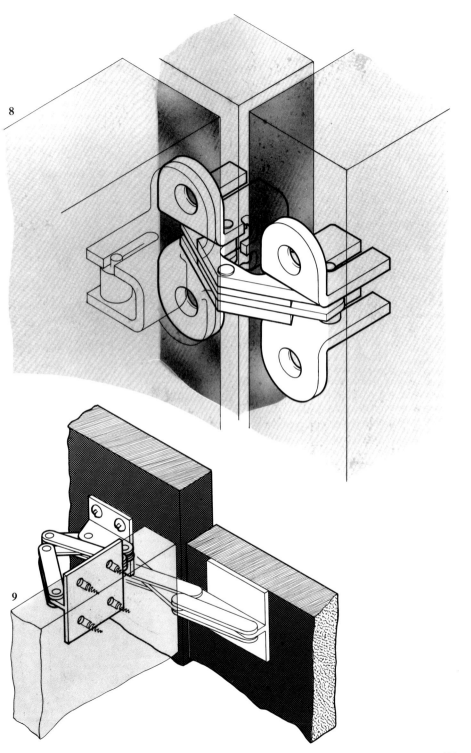

8

9

77

Other hinges

A similar situation arises on modern kitchen cupboards where the doors are all aligned and when opened must move outwards to avoid fouling the adjacent doors, then rotate round to end fully open in front of the original door line. Various mechanisms have been devised to do this and one is shown in figure **10**.

With the advent of cheap aluminium extrusions a completely new type of hinge has recently been introduced. The Rotalex Continuous Hinge (figure **11**) can be as long as the door or flap to which it is fitted. With all-aluminium construction it can actually be extruded as part of the door and door frame, needing only the small channel to complete the hinge. Small plastics blocks are inserted at intervals to take the longitudinal loads.

This quick review of the basic types of hinge is not exhaustive. There are many variations on each type and many more types than are mentioned here. Remember that the objective is to study the mechanism involved. Investigate as many types as you can find and analyse their movements and their load paths for various door positions.

As a follow-up exercise try to design a hinge mechanism for an aircraft door. The door and its frame are curved and no part of the mechanism may protrude beyond the body profile. The mechanism must not take up too much room inside the aircraft or make it difficult to open and close the door. The door must withstand internal pressure loads and as it closes it should move inwards to press upon a seal without pinching or stretching it. When the door is open it should not obstruct the opening or block access.

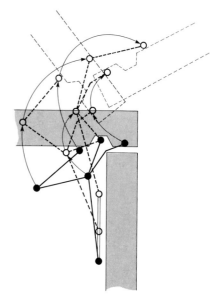

10 *This kitchen cabinet hinge allows the door to clear adjacent doors*

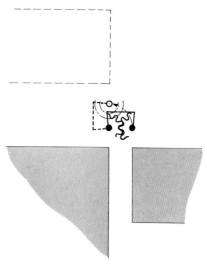

11 *The Rotalex Continuous Hinge uses extruded gears to provide the necessary motion*

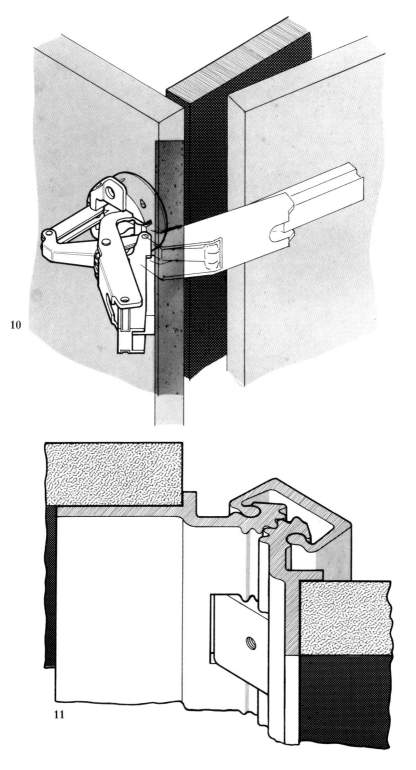

10

11

Folding structures

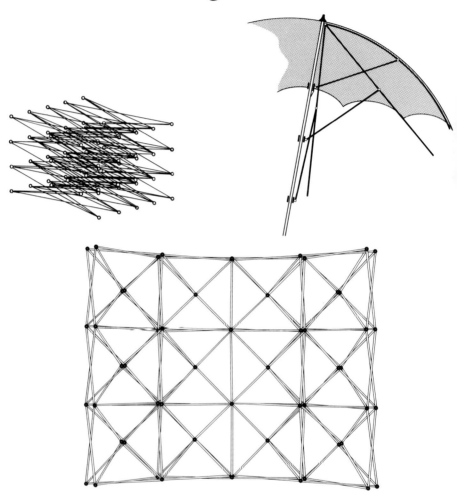

There is frequently a need for a structure to support something for a particular purpose, but for various reasons a permanent structure is not appropriate. Hence a structure which folds away is designed. There are three basic types:

1 Structures which fold to form a package for storage or transport.

2 Adjustable structures which may be set up to support loads in a variety of working positions.

3 Structures which continue to support loads while movement takes place, without doing significant mechanical work. If work took place the device would be strictly a mechanism. This ignores work against friction forces and transient work involved with dynamic loads.

In addition the device must retain sufficient internal structural integrity of parts. Thus standard scaffolding does not form a folding structure but a temporary

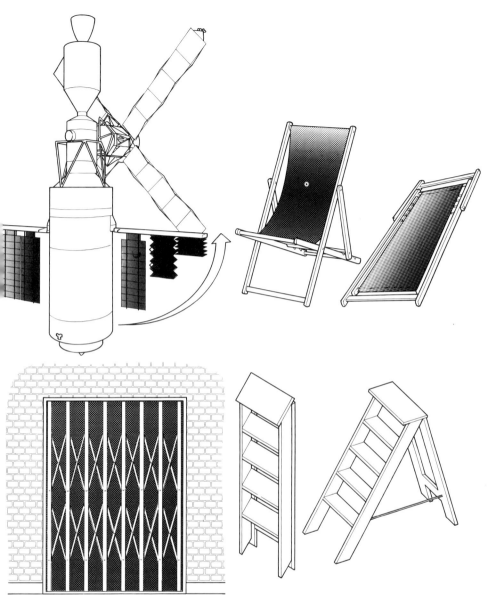

structure. By convention chains, ropes, cloth, or other one- or two-dimensional tension-resisting structures are not considered as folding structures but have their own classifications. Folding structures usually resist bending, torsion, or compression loads or any combination thereof, and perhaps tension as well.

Fold–away structures

These are mechanisms which lock in a particular configuration to become structures. Everyday examples include umbrellas, step-ladders, sun awnings (see page 65), and folding tables and chairs. Engineering examples are swing or lift bridges, the Thames Barrier, thrust reversers on jet engines, and solar arrays on spacecraft. Even simple doors may be classified in this way.

Adjustable structures

These are mechanisms which may be locked in at least two positions to support loads, usually the same type of load but sometimes a completely different load. Common examples are extension ladders, adjustable shelving, and most car seats. Each of these may be locked in a number of positions. Historically, castle drawbridges were load-carrying beams in their open position, but when raised they withstood the shock loads of battering rams and missiles. In engineering many devices are adjustable in stages. The simple plastic cable tie with a ratchet grooved face running past a gripper is one of the smallest. Others include railway points and jacks used by builders where a short-travel screw jack can be repositioned by a pin-in-hole arrangement. Some devices use friction grips to lock hinges or sliders in any position thus providing a structure which is infinitely adjustable, but its load-carrying capabilities are limited by the friction available and they should be used with caution.

Movable loads

The borderline between structures which support loads while they are changing shape and load-carrying mechanisms is very hazy. The same device may serve both purposes depending on circumstances. Folding brackets to support televisions allow horizontal movement while resisting vertical loads. This is typical of this type of structure. It is a structure in some planes and a mechanism in others. A drawer is a good example, whether on simple slides or the telescoping roller systems in filing cabinets. The load in the drawer is supported laterally and vertically at all times, but may move easily back and forth. The distinction between this sort of device and any other tracked system is the property of extending and retracting rather than simply transporting over a fixed support.

Designing folding structures

As with other mechanisms the components
of folding structures are connected by pivots,
with one, two, or three axes, or sliders.
Many types of mechanism can form folding
structures but certain ones are particularly of
note. The simple hinge where one part folds
against another to form a more compact
arrangement is almost a trivial example, but
in multiples it forms the zig-zag of fan folds
used in various places. A pair of such hinges
pivoted to each other forms a scissors
arrangement and multiples of this produce
lazy-tong mechanisms which can be locked
to form structures. Simple slides connecting
parallel components also provide a means of
folding a structure, typified by the telescope.
Multiples of these provide greater degrees of
extension.

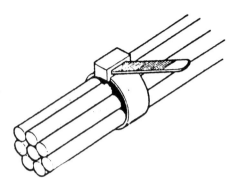

For mechanical joints to carry structural
loads the areas of contact between
components must be arranged to transmit
the stresses involved and yet maintain
sufficient clearance to allow movement. This
inevitably means that the structure has a
certain amount of slack. This may lead to
some distortion under load which can be
quite acceptable. However, if the load
changes direction, the joint slack may allow
movement and consequent dynamic and
impact loadings within the joints, thus
causing structural failure from fatigue or
other effects. These problems can be
alleviated by using long pivot pins of large
diameter and long overlaps on sliding parts.
These reduce the slack in the joints and also
reduce the stress levels necessary to provide
bending connections.

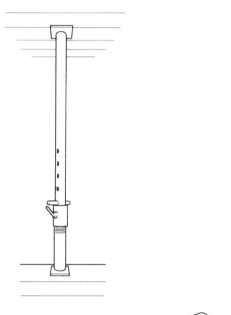

Keyboards

A keyboard is a series of keys arranged in a more-or-less conventional pattern. The most common type is the QWERTY used on typewriters and computer terminals throughout the world. The name is derived from the first six keys on the top line. There are many other types of keyboard: from typewriters to telephones, from cash registers to calculators, and from musical instruments to custom-designed switchboards. The technology used in their manufacture can vary from mechanical linkages to the latest microprocessor-controlled membrane switches with tactile and audio feedback.

Let us start with a brief history of the development of the QWERTY-type keyboard. The first typewriters were coming into production during the Victorian era, but there was no agreement as to the order in which keys should be placed.

Initially it was alphabetical order. Soon, however, it was found that some of the more common letters, 'e' for example, were in positions that ensured the imprint on the paper would be very heavy, whereas others such as 'z' were in positions such that the final imprint was very faint.

It was discovered that the factors that influenced the boldness of the imprint were, among others, the strength of the finger that struck the key, the frequency of use, and the length of the lever arm between key and striking hammer.

The key for letter 'e' for example (in the position now occupied by 't'), would be struck by one of the strongest fingers on the right hand and this is the most commonly used letter in the English language. Furthermore, the lever arm for 'e' was very short. Together these factors led to a very strong imprint being left on the paper and

1 *Remington Model 2 in use around 1888*

84

increased wear on the hammer. In the case of the letter 'z' the opposite applied: it is used only rarely and, being positioned at the extreme end of the keyboard, its lever arm was long and it was struck with one of the weakest fingers. The imprint, therefore, was very weak.

For this and other reasons the layout of the keyboards was redesigned to produce a more even line of type and the QWERTY arrangement became the accepted standard. Thus even though in the age of electronic keyboards and computers these constraints no longer apply, the layout of the keys remains. To an outsider, or someone who has not seen a QWERTY-type keyboard before, the arrangement might be seen as being deliberately designed to slow down the process of inscribing words on to paper!

The designer of a new keyboard has to decide whether to abandon this convention and produce a novel, if more logical, arrangement of keys, or to go with tradition and the thousands of touch-typists who have learned to use the standard keyboard and who would be lost on a new design. In the event practically all keyboards that have a separate key for each letter do follow this traditional QWERTY pattern. There are significant differences, though, in other ergonomic factors such as the pressure required to make and break the switch, the pitch of the keys, and the rake or steepness of the lines of keys. Some keyboards are now available that are 'human factors engineeered' or designed specifically to reduce the effort required of the typist. This is achieved by contouring the keyboard to the shape of a pair of hands and reducing the number of awkward moves that have to be made to reach the less common keys.

2 *Example of an ergonomic keyboard design*

Other types of keyboard

This realisation, that electronic keyboards do not have to fit in the same mould as the previous mechanical ones, has started to lead to new and innovative designs, particularly in the form of coded-entry keyboards. This type of system uses a relatively small number of keys, usually ten or less, any of which may be depressed simultaneously so that a particular combination of keys corresponds to a unique character. With ten keys there are over 1000 possible different combinations. This is easily sufficient to cover all the characters that fit into the standard electronic code. Known as ASCII (American Standard Code for Information Interchange) and pronounced 'ass-key', this is an almost industry-wide standard whereby each letter or character is represented by a binary number. There are 255 different characters available and this accommodates all the alphanumeric characters of most major European languages, as well as all the specialised word processing commands required of a so-called 'intelligent keyboard'.

The manufacturers of coded-entry keyboards claim that the code is very swiftly learned and suggest that after only a few days practice it is typical for an initiate to be able to 'write' more quickly with it than in longhand script. These new keyboards are designed to be compatible with computers and word processors and can usually store more than four A4 pages of script.

The Microwriter is a six-key coded-entry keyboard. The keys are not all arranged in a straight line and this aids the recollection of the letter codes. If the spatial relationship of the keys is kept in mind, the shape described by depressing the thumb, the first finger, and the little finger is that of an 'L'. This is, in fact, the key code for the letter L. Other letters are remembered in a similar manner. It is this method that allows the code to be memorised easily and permits high microwriting speeds to be attained in a relatively short period of time.

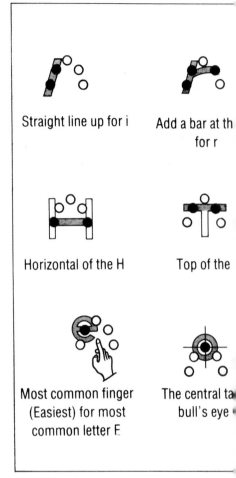

Straight line up for i

Add a bar at th for r

Horizontal of the H

Top of the

Most common finger (Easiest) for most common letter E

The central ta bull's eye

3 *Examples of coded entries*

A coded-entry keyboard is therefore ideally suited to electronic data entry, whether into a solid-state memory, or directly on to paper.

This leads on to new developments in cash registers, or point of sale terminals as they are becoming known. In many fast-food restaurants where there is only a limited choice of items for sale, or when there are only a few basic meals, each with a limited

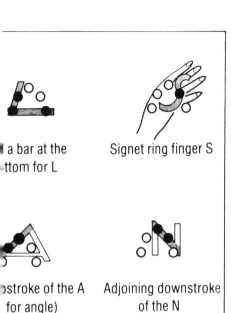

a bar at the
ttom for L

Signet ring finger S

)stroke of the A
(for angle)

Adjoining downstroke
of the N

Space

4 *Microwriter in use*

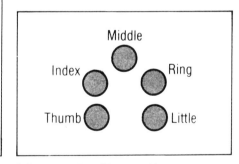

5 *Representation of finger positions*

number of variations, it is possible to use a new type of cash register which has many more keys than the older mechanical types; in fact there could be as many as one key per item on the menu.

For an order of $\frac{1}{4}$lb cheeseburger with large fries and a small cola the assistant would press the three appropriate buttons, the receipt would then be printed and the total price shown on the display. In this way the assistant does not have to remember the prices of all the various meals and options. Thus both the operator training time and customer delay are reduced so increasing the productivity of each point of sale.

These new cash registers are programmable so that if the prices change, or the selection of goods for sale alters, a few small changes to the software will ensure that the terminal is kept up to date. They can also be linked to a central computer that automatically keeps a record of the type of goods that are sold, at what time of day, and which day of the week, so that the manager can analyse his sales and order accordingly in the future.

Because many of these point of sale terminals (the distinction lies in the ability to communicate sales information to a central computer) are used in the food industry, they are made more hygienic by the use of new membrane keyboards. Membrane keyboards consist of three basic parts: the membrane, a spacer, and the substrate. The membrane is a polymeric film, usually polyester, which has flexible conductors attached to its lower face in positions that correspond to switch locations. The upper surface of the membrane is often bonded to a further panel with such symbols, colours, and instructions as may be required in terms of graphic design.

The spacer is an insulating layer that separates the membrane from the substrate by between 0.1 and 0.2mm. The substrate is either a flexible or rigid printed-circuit board. The flexible membrane, with conductive material on its lower side, is separated from the substrate, with switch contacts on its upper surface, by the thin spacer. This spacer has holes in positions corresponding to the switch positions on the graphics panel.

When one of the switches is depressed the conductive surface of the upper membrane touches the contact points below it and acts as a shorting bar to close the circuit. When the switch is subsequently released the resilient membrane breaks contact and returns to its original position. The life expectancy of these switches is of the order of 10 million cycles, although the maximum current capability is not more than a few tens of milliamps.

In this manner a keyboard that is only a few millimetres thick, custom-designed in terms of layout and graphic design and hygienically sealed against undesirable dirt

particles can be produced for a fraction of the cost of a conventional push-button arrangement.

The biggest difficulty to be overcome when using the basic membrane keyboard is the lack of feedback to the operator. Because the travel of the key is so small, the operator is not directly aware that the switch has made contact. There are a number of ways of overcoming this challenge. The cheapest way is to use audio feedback, whereby a buzzer or whistle is made to sound whenever a key is depressed. This can be ineffectual, however, in a noisy environment or offensive when the surroundings are quiet. The manufacturers of cash-dispensing machines have observed this: in busy streets there are more miss-keys than in quieter locations.

Another way of providing feedback is through the sense of touch: 'tactile feedback'. The most cost-effective way of obtaining this (for mass-production) is to use the standard membrane keyboard but have a custom-built rubber overlay with buttons in appropriate places. These buttons are so designed that when a certain force is applied they buckle inwards and transmit the force on to the membrane keyboard below. The buckling force is greater than the required contact force so the operator knows that the switch has been made when the movement of the buckling rubber is felt. Other ways of overcoming this problem are being discovered all the time and there are a number of patented solutions available.

The chief advantages of membrane keyboards compared with push-button types, are therefore low cost, low breakdown rate, and simplicity.

Of the predominantly numeric keyboards that are available the most ubiquitous must be the pocket calculator. Again the early varieties did not make provision for feedback to the operator. Electronically generated audio feedback is out of the question (think of an examination room!). Despite this constraint most manufacturers

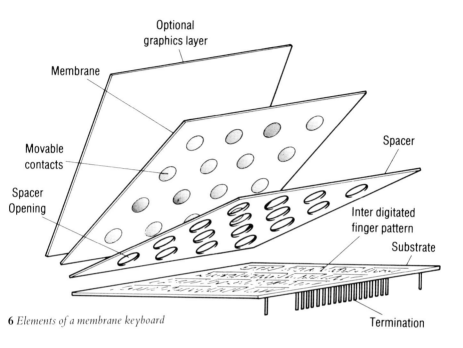

Optional
graphics layer

Membrane

Movable
contacts

Spacer
Opening

Spacer

Inter digitated
finger pattern

Substrate

6 *Elements of a membrane keyboard*

Termination

seem to have come to appreciate the importance of incorporating some feedback in their products and now use either a soft click or a dual-pressure type system whereby a light touch gives a small degree of movement, but a slightly heavier push will depress the key completely.

An interesting observation is the difference in layout between calculator keyboards and telephone key pads. Calculators seem to have standardised their layout with 'O' at the bottom and the remaining numerals working from left to right *up* the keyboard. Telephones, on the other hand, have the numerals 1–9 arranged from left to right, but reading *downwards* again with 'O' at the bottom.

Telephones do not have any direct feedback system, although the digital

(computerised) exchanges that are common in the United States recognise only tonal inputs as opposed to the series of clicks that characterises the electromagnetic exchanges predominant in this country. Thus with the digital system as each key is pressed a tone of set frequency and interval is sent to both the exchange and the earpiece of the handset. This tone provides the user with ample feedback when accompanied by the movement of the key as it is depressed.

The last big category of keyboards is the musical type. The electronic keyboards now coming out of Japan and Hong Kong in such profusion are again just a selection of switches, admittedly with some very sophisticated electronics to back them up. Nevertheless the standard piano or organ type of keyboard is composed of a series of switches that tell the internal processor which tone or chord to generate.

The original musical instruments from which these keyboards have developed had mechanical linkages between the keys and the actuators (hammers and air valves in pianos and organs or quills in harpsichords). In this sense they paralleled the original typewriters that were mentioned earlier. The expensive synthesisers favoured by the big electronic rock groups of the 1970s could be compared with the specialised data processors of the same period and the current mass-produced low-priced synthesisers that are now coming out of the Far East could be compared with the word processors or programmable calculators in other fields.

This growth in the markets of all these three sectors – calculators, word processors, and musical synthesisers – is a direct visible result of the low cost and versatility of the microprocessor chip. All these products require keyboards of one type or another and therefore this market too will be seen to expand significantly in the next few years; perhaps more new technologies will also be seen to emerge. In fact one of the few external signs on equipment that contains such microprocessors, is the appearance of a keyboard on the front panel. Take coffee machines; these can now offer options such as extra milk or extra sugar where once one was restricted to just 'with or without'. The selection keys have changed from the heavy electro-mechanical type that might be switching mains voltages and high currents to electronic or solid-state types that send a signal of a few volts at milliamp levels to a central controller which then initiates the mechanical sequence of events by means of relays or other actuators. The controller or processor can then keep track of the sequence and is more flexible in the case of a malfunction. Drink selection is by keying a number that represents a certain combination of milk, sugar, coffee, etc.

Thus a seemingly simple topic such as keyboards has many design implications –

ergonomics, cost, reliability, industry standards, government standards, market research, etc – as well as the engineering difficulties of introducing tactile feedback to a custom-designed membrane keyboard, for example, and then interfacing the keyboard with a central computer.

It is instructive to observe how many different types of keyboard you use, how many you see others using, and which technologies are used in which of the different types. This applies by extension to all aspects of engineering design: detailed observation of accepted designs will help the engineer to produce new designs that will in their turn be acceptable or perhaps to use an idea from one field (calculators) in another, superficially different, field (musical instruments).

7 *Typical keyboard application*

Can openers

In engineering terms a can opener is a metal piercing and cutting device for use on thin sheet metal in a limited range of configurations by unskilled operators in a hygienic environment. An examination of the history and development of the various types of can opener serves as a useful pattern for the examination of other kinds of hand tool, and as an indication of how different designers find different solutions to the same problem. It is not the intention to examine the ergonomic or aesthetic design problems here, but only to look at the functional solutions to the problem of opening a can.

Historical notes

During the Napoleonic Wars the problem of maintaining ships in fighting trim for extended periods at sea, a problem not completely solved today, was compounded by the difficulty of supplying fresh food to patrolling men-of-war. Drying, salting, and pickling helped to preserve some of the qualities of meat and vegetables, but deficiencies of vitamins and other essentials

soon led to scurvy and other conditions which sapped the strength, and often took the lives, of crews. Two different but similar inventions to help in food preservation appeared at this time. In France in 1809 Nicolas Appert received 12,000 francs from Napoleon for the process of heating food and sealing it in glass containers. This was known in the United Kingdom as bottling, but, confusingly, was called canning in the United States. It was particularly good for fruit and vegetables, but the bottles were rather fragile and were not used extensively by the French Navy. In England a similar process was invented but this time using heavy metal containers sealed with molten lead. The resulting cans were robust and were soon approved by their Lordships at the Admiralty thus making some contribution to the defeat of the French Navy by Nelson. There was at that time no special tool for opening cans and a hammer and chisel were used (figure **1**). A surviving sealed can from that time was recently opened and its contents found to be quite fresh.

1

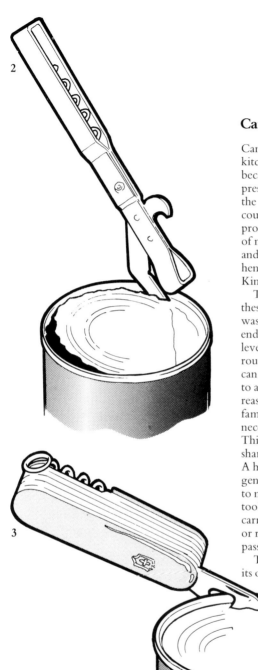

Can cutters

Cans did not make a great impact on the kitchens of the world until thin sheet steel became readily available and rolling, pressing, and crimping were developed to the stage where lightweight cylindrical cans could be produced. At the same time the process of tin plating eliminated problems of metal contamination, particularly by lead, and gave us the modern type of tinned can, hence 'tin can' or just 'tin' used in the United Kingdom.

The time was now ripe for a tool to open these cans and the spiked-blade lever type was born. The spike is used to pierce the can end with a stabbing motion and then a levering procedure forces the blade to cut round the disc, using the rolled edge of the can on which to rest an extension of the tool to act as a fulcrum (figure **2**). (I have no reason to suspect that the reader is not familiar with this, but a description is necessary for comparison with later types.) This is essentially a cutting process with a sharp blade being forced through the metal. A high mechanical advantage is necessary to generate the required force. It is interesting to note that a very similar but much larger tool is included in the emergency toolkit carried on all trains to cut through the sides or roof of a carriage to rescue trapped passengers after an accident.

This early design is still available in almost its original form as well as in a number of minor variations, particularly folding versions as part of other kitchen tools or penknives (figure **3**). Its main disadvantages are the dangers arising from the combination of a sharp point and the difficulty of holding a round tin. Moreover, the jagged edge produced is not only dangerous but can make it difficult to remove the can's contents.

There are a number of improvements on this cutter type. One simple folding version is issued to the armed forces and works surprisingly well. A small folding blade is hinged to a simple flat plate with a notch in it. The blade is hooked over the can rim and the notch hooks under it (figure **4**). A twisting motion then forces the blade into the can making a short cut. The process is repeated, cutting round the can end. The tool is slow and requires some effort to use, but most of the earlier disadvantages are eliminated although the cut edge is still a little ragged.

A modern can-cutter for the kitchen is shown in figure **5**. The small blade is driven into the can by a knock from the heel of the hand, or something harder if necessary, and then a to-and-fro movement of the handle causes the toothed wheel to bear against the can rim pulling the blade round the can top. The angled teeth give the wheel a ratchet action moving the blade in one direction. The tool is easy to use and produces a smooth even cut.

4

5

Shearing can openers

This type of opener has largely displaced the older cutters in the domestic kitchen. They are easier to operate, requiring less effort, and leave a smooth edge to the can. However, they often have plastic parts or require oil on bearings and have awkward corners in the most critical areas, making many of them difficult to clean and impossible to sterilise. This reduces the level of hygiene which can be maintained and is a potential source of food poisoning.

6

The basic version of this type of tool has a small shearing-blade with a pointed end and carries a grooved drive wheel and a butterfly knob. The drive wheel engages the underside of the rim and when the handles are squeezed the point of the blade pierces the can (figure **6**). Maintaining pressure on the handles with one hand, the butterfly knob rotates the drive wheel to push the shearing-blade round the rim of the can. Left-handed versions are available.

7

Variations on this basic type include replacement of the fixed blade with a wheel which rotates giving a truer shearing action, through the metal rather than along it. This makes the butterfly handle easier to turn and gives a longer life to the cutting edge. The drive wheel and shear wheel are often geared together rather than relying on friction to turn the cutting edge (figure **7**). An added refinement is a magnetic lid-lifter to stop the

8

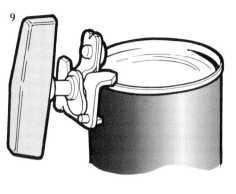

9

severed can end sinking into your soup. The use of these magnetic devices, particularly on electric can-openers, has meant that modern lightweight aluminium cans have to have a steel end cap or the lid-lifter would not work.

There are a number of designs of shear openers which remove the need for squeeze handles. One of the simplest is in figure **8**. When placed on the rim of the can the initial turn of the drive wheel forces the blade point by lever action through the can and locks it into its cutting position. Further rotation then moves the tool round the can slicing off the end. Other varieties use a cam (figure **9**) or a toothed rack (figure **10**), to achieve the initial piercing. A slightly different approach turns the whole mechanism round so that the drive wheel rests above the rim and the shear wheel cuts through the side of the can, removing the whole can end including the rim (figure **11**). This is a neat solution but leaves a circular sharp edge exposed on the can.

The cutting ability of all shear types is very sensitive to the relative positions of the drive wheel and blade. If they are too close, the drive wheel rides up a narrow can rim and the tool jumps off the can. If they are too far apart, there is insufficient grip to drive the mechanism. Some versions have strong springs designed to maintain the correct relationship over the range of possible rim sizes.

10

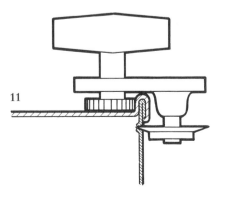

11

Peel-off lids

The forerunner of this type of can-opener is the key-opened sardine can. One side of the can is sealed in place with a deliberately weak solder. Since there is little pressure-loading, the join is strong enough in total to maintain the seal. However, the joint does not have a high peel strength and, using a simple key on a tab as a tool, the whole can side may be peeled off (figure **12**). Similar key systems use weakening grooves in the can side to produce a tear strip around the can, typically for corned beef. When metal-pressing techniques allowed weak tear grooves to be pressed into can-end material, the familiar ring-pull cans were introduced (figure **13**). Soft drinks and beer cans with a limited tear-out region are well known, but a number of cans are now appearing where the entire can end tears off, usually in a spiral strip.

Miscellaneous

There are various can piercers for liquid contents which simply make a hole to allow pouring. They are usually simple spike or point tools which pivot on the can rim (figure **14**). An alternative is the adhesive tear-off strip used mainly on large beer cans.

Notes

Opening a can may seem a simple task, but the various solutions examined here have illustrated a number of fundamental principles. The difference between cutting and shearing, levers, friction drives, gears, material properties, and, of course, design ingenuity may all be examined in these everyday objects. The variety shown here is not exhaustive and there are many minor variations not shown.

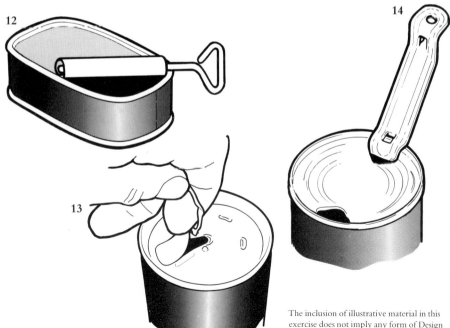

12

13

14

The inclusion of illustrative material in this exercise does not imply any form of Design Council approval for the products shown.

Springs

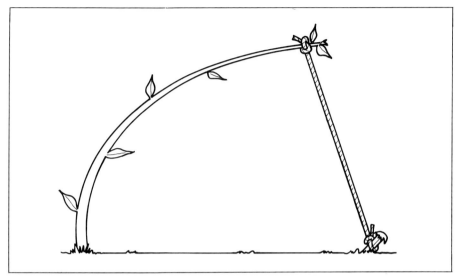

1 *One of the earliest energy storage systems*

All solid materials, and structures made from them, deflect when forces are applied to them. This is a necessary part of the way in which solids retain their shape; a system of forces must be generated within the solid to exactly balance the externally applied forces. These internal forces arise when the atomic bonds are pushed away from their equilibrium position. The total effect of the bond distortion is an overall deflection of the solid.

Most structures are designed to keep deflections small. Indeed the equations used in most circumstances assume that deflection is so small that the basic geometry is unchanged. However, if a designer wishes to make use of the energy stored in a structure, very small deflections, even for large variations in loads, are not very convenient. To overcome this problem a number of structural shapes have been conceived with different characteristics of deflection under load. These are collectively known as springs.

Beams as springs

Despite being mechanical devices associated with movement and energy storage, springs are structures and the analysis of spring behaviour and their load and deflection limitations is a structural problem. The earliest use of springs was probably bent saplings or tree branches used to 'spring' animal traps. Such springs are long slender beams firmly anchored at one end to form a cantilever. The free end has a considerable range of movement for a modest change in load. More importantly the stress levels within the wood remain low enough to avoid breaking it. The natural taper of a tree or branch is significant here.

The bow illustrates this point more effectively. It is a double cantilever beam used as a spring. The highest bending loads occur in the middle at the handgrip and the wood is thick to cope with them. If the whole length of the bow were this thick the bow would be too stiff; the deflection under

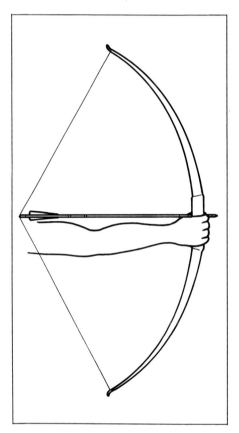

2 *The bent sapling principle was developed into an energy conversion system*

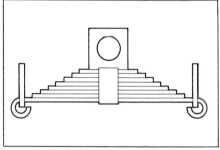

3 *The tapering and stiffness are still key elements in the design of leaf springs*

load would be small and the amount of energy available to propel the arrow would be insufficient. However, the bow tapers towards the ends as the bending loads decrease; this keeps the stress levels higher. Hence the deflections are quite significant and the stored energy, the integral of force × distance, is quite high. A useful secondary effect, in what is a dynamic application, is the low mass and hence low inertia of the tapered spring. Thus the available force is not wasted accelerating the spring.

Cantilever springs are still used in their simple form in some railway and truck suspensions, usually as a laminated beam, but new materials and techniques are allowing one-piece tapered springs to achieve the high performances required. They also appear as swimming pool spring-boards and provide the small forces required in some locks and other small mechanisms. However, for most purposes they have been superseded by more complex but compact configurations.

Helical springs

Perhaps the most universally known spring is the helical spring. This is formed by coiling a soft wire round a former, then heat treating it to harden and stiffen it in its helical shape. Normally used in tension or compression along its axis, the coiled shape distorts the wire by twisting it. The torsion stresses are at the same level at all points on the wire for a uniform cylindrical helical spring, except locally at the ends where the loads are applied. In this simple form the helical spring is an inexpensive and compact way of storing energy and balancing loads. It has a multitude of applications from retractable pens to car suspensions.

Tension springs may be of any length, but require hooks or other end connections to transfer loads. Compression springs, like other compression members, are limited in length by their tendency to buckle. Where longer springs are needed some form of guidance is necessary. The guides must be arranged to allow for the slight increase in overall diameter of the coils as the spring compresses. Not all compression springs are of constant diameter, some taper along their length or have other shapes for specific applications. One advantage of a tapered spring is that it can be compressed much more; the coils lie inside each other and, if the material allows, the fully compressed height may be the diameter of the wire.

Torsion springs

The other major type of spring is the torsion spring. This gives a rotational torque rather than the straight force available from the springs described previously. An early form of torsion spring is again a straight bar clamped at one end and loaded at the other, but this time the load is applied as a twist on the bar. It may be formed as a door-closing spring lying along the line of the hinges, with one end fastened to the frame and the

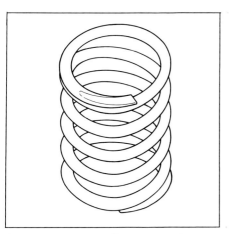

4 *Improvements in metallurgy permitted more compact springs*

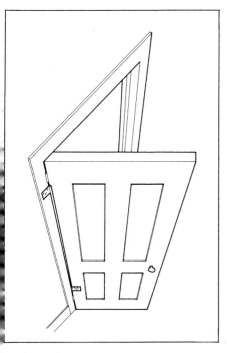

5 *The simplest torsion springs are in some ways analogous to beams in simple bending*

other to the door. Torsion bars or tubes are frequently part of road and rail suspensions and appear in many other mechanisms. The loading is usually applied via a lever arm rotating about the bar centre line.

The other major type of torsion spring is a flat coil of wire or strip material. In an open coil, where the parts do not touch each other, the spring is used in lightly loaded dynamic applications, such as the hairspring on a (mechanical) watch. In a close-coiled form the spring is used as a quasistatic power store, such as the driving spring on a watch. Spring motors using this latter form may be wound up from the inside, as in a watch, or from the reversal of the mechanism, as in seat-belt retractors and push-pull rules. One major advantage of such a power spring is the considerable amount of rotation which can be obtained for a relatively small variation in force, ie the low spring rate.

An interesting modification of the power spring is the constant-force or zero-rate spring known in the United Kingdom as the Tensator. Here the spring itself is pulled off its coil, which is free to rotate. The straightened portion has little effect and the force is generated by the curved sections. The lengths and curvature of these sections remain constant for most of the positions of the spring and hence the force is constant. These springs are used to counterbalance the weight of machine tools, drop windows, and similar applications eliminating the need for counterweights and pulleys. If the end of

such a spring is wound round a shaft it produces a constant-torque motor. The close-coiled helical spring can also be used as a torsion spring. The ends of the wire are usually formed into shapes which allow the torsion loads to be applied.

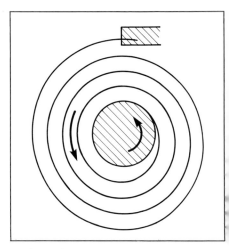

7 *Close-coiled torsion springs are analogous to helical springs in compactness and energy storage capacity*

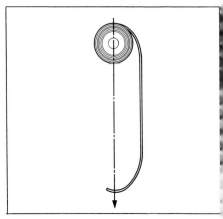

8 *The Tensator spring is a specialised application of the torsion spring*

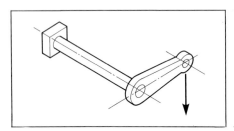

6 *This type is frequently used in high-load applications*

Disc springs

So far the springs described have been made from wire, usually round but sometimes rectangular; the extreme case being the narrow strips used for the torsion power springs. This section covers coned disc springs sometimes called Belleville washers, but this name applies only to a specific ratio of dimensions. The coned disc spring is made from a flat disc with a central hole which is formed into a conical shape. These useful little springs may be used singly or stacked in series or parallel configurations to give a wide range of force/deflection characteristics. The diaphragm spring is a coned disc spring with a series of slots cut in the inner edge. This gives a 'flat' behaviour more useful in certain applications such as automobile clutches.

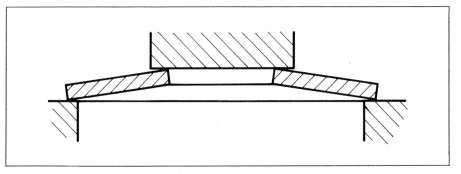

9 *Disc springs may be used singly . . .*

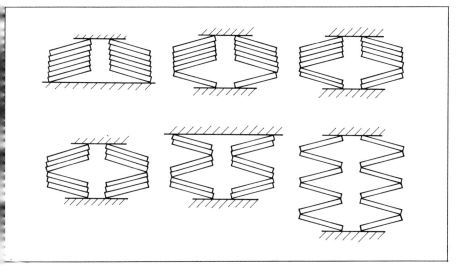

10 *. . . or in series or parallel or any combination*

Ring springs

It takes a very high loading to expand or compress a heavy steel ring. However, if a series of rings bear against each other on steep wedges arranged to expand or compress alternate rings, the necessary loads can be generated. The amount of energy that can be stored in such a ring stack is very high indeed. The main use of such springs is in railway buffers where the high friction between rings helps to dissipate as much as 60% of the energy although plastic deformation may occur.

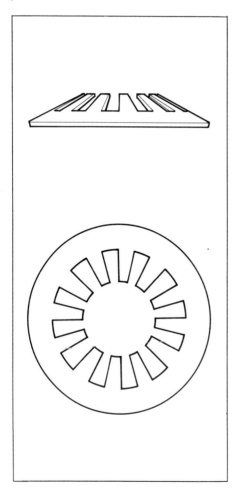

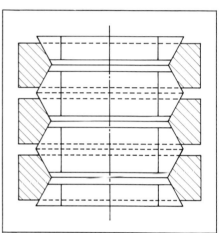

12 *If energy absorption is a problem, the railway buffer provides a heavyweight solution*

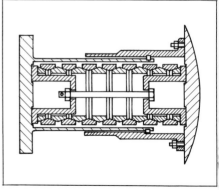

11 *The diaphragm spring is a variation that gives a more linear response*

13 *The central bolt should only be released with great caution!*

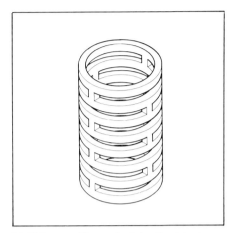

14 *This type of flexure lends itself to injection moulding techniques*

15 *The new sulcated spring is also made from plastics and looks set to find a large market in the vehicle industry*

Unusual springs

Various other spring configurations have been proposed, but rarely do they prove cheap enough or have some special characteristic which lets them replace some of the main spring types described above. In the early 1960s a very simple type of spring was proposed; made by simply cutting a series of slots in a tube. This effectively produced a series of rings loaded in opposite directions at $90°$ intervals. The behaviour of these springs can be tailored by varying the width and spacing of the slots. They can be made in many materials and usually require no heat treatment after machining. Their behaviour range falls between that of helical springs and conical disc springs and has some advantages over both. However, it has not found a large market.

Recently the sulcated spring received the 1983 British Innovation Award. This spring is made from four corrugated plastic arms linking two end plates. The concave areas are filled with an energy-absorbing plastic. The assembly is designed to replace a spring/damper combination on a car suspension and is approximately half the weight for the same duty. It has useful lateral stiffness and will not corrode. It is also claimed to be fail safe since the breakage of one arm would still leave a significant load-carrying capacity.

Other springs

There are many other types of spring including gas springs and rubber springs. They are often used in combination with each other with pre-loading and in series and parallel. But no simple introduction could possibly cover them all.

Further reading

Engineering Design Guides, published by Oxford University Press:
08 Helical springs
42 Mechanical springs

Cork extraction

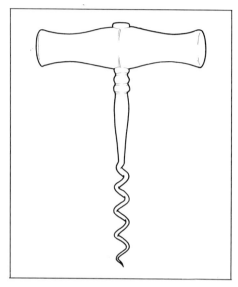

1 *A regular, uncomplicated corkscrew*

This unit investigates ways of removing the cork from a bottle of wine. The requirement may be easily understood. The neck of a wine bottle is sealed with a tight-fitting cylindrical plug of cork, which is flush with, or slightly inside, the top of the neck. In order to drink the wine you must first remove the cork.

If you have an appropriately shaped rod (or even a strong thumb!), you can usually push the cork down into the bottle. This is not recommended, since wine tends to spray out and the floating cork makes pouring difficult. In the absence of any other tool an old book on wine recommends heating the neck of the bottle in a candle flame, then drawing a line around the neck just below the cork level with a wet feather. With luck the local cooling effect causes the glass to crack and the plugged neck can be removed. This is a time-consuming and potentially dangerous activity but may be appropriate on occasion.

Gripping the cork

The first problem is to get some purchase on the cork. It is desirable to cause the minimum of damage to the cork to avoid dropping bits of it into the wine. The most frequently used solution is the corkscrew. This is usually a piece of thick wire bent into a helix and sharpened at the end. The helix is usually right-handed, but some left-handed versions are available. A variation on this is a straight rod with a helical flute resembling a gimlet boring-tool. This is stronger than the helical wire but more likely to damage the cork. These devices are simply screwed into the cork.

A second solution employs a tool with two thin flat prongs which are pushed down on opposite sides of the neck between the cork and the glass. A rotation of the tool breaks the long-term seal and allows the cork to be eased out. This device is successful in most cases, but can fail with very tight corks.

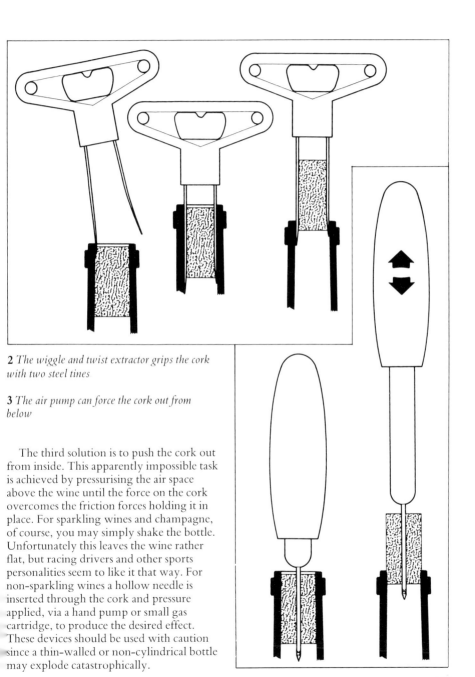

2 *The wiggle and twist extractor grips the cork with two steel tines*

3 *The air pump can force the cork out from below*

The third solution is to push the cork out from inside. This apparently impossible task is achieved by pressurising the air space above the wine until the force on the cork overcomes the friction forces holding it in place. For sparkling wines and champagne, of course, you may simply shake the bottle. Unfortunately this leaves the wine rather flat, but racing drivers and other sports personalities seem to like it that way. For non-sparkling wines a hollow needle is inserted through the cork and pressure applied, via a hand pump or small gas cartridge, to produce the desired effect. These devices should be used with caution since a thin-walled or non-cylindrical bottle may explode catastrophically.

Exerting a pull

Once your corkscrew is screwed well into
the cork, there is then the problem of
exerting enough force to separate cork from
bottle. The simplest corkscrew just has a
handle; you supply the pull. If you happen
to be fairly strong you should be able to
manage most corks, but some will need a lot
of sweat and bother; hardly the right activity
for your quiet dinner party.

If your muscles are not in peak condition
or you prefer a more delicate approach,
some device to increase your leverage is
obviously required. These tend to fall into
two main categories: levers and screws. Both
must pull on the cork and push on the bottle
to extract the cork.

Levers

The simplest of these is frequently used by
wine waiters since it folds to fit neatly in the
pocket. It has a corkscrew centrally placed
on a flat bar with a hinged piece at one end
shaped to fit on the lip of the bottle. Pulling
on the other end gives a leverage of about
2:1, which will shift most corks. A little care
is needed to hold the hinged piece in
position, but it performs very well.

4 The waiter's friend acts as a simple lever

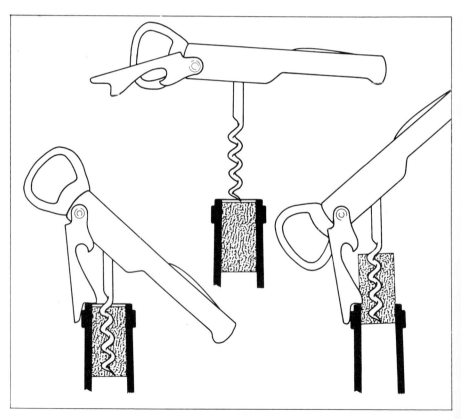

A popular device has two side levers. The top of the corkscrew has a rod with a series of rings forming a rack for a pair of quadrant gears each part of a short lever. The two levers are carried in a frame surrounding the rod, with a ring at the bottom to fit the neck of the bottle. As the corkscrew goes into the cork the frame meets the bottle neck and the movement of the neck causes the levers to lift. If both handles are pressed down a leverage of about 4:1 easily extracts the cork.

5 *Levers are also used in other devices*

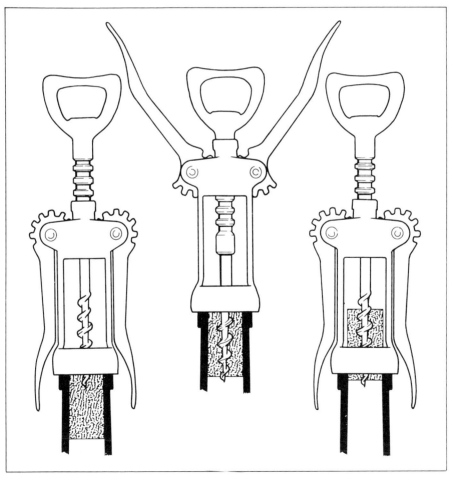

A French corkscrew uses a lazy tongs action to give an even larger lever arm. A four-stage lazy tongs has a neck ring at its end crossing and a corkscrew on the next crossing. A spring keeps the levers together as the corkscrew is inserted. Once the neck ring has contacted the bottle a steady pull on the handle moves the most difficult corks with a lever ratio of about 4:1. The device is a little awkward to screw into the cork, but very satisfying in its pulling action. Since the pull is in line with the cork the effort needed is reduced.

6 The lazy-tongs gives a lever ratio of 4:1, but is not very common

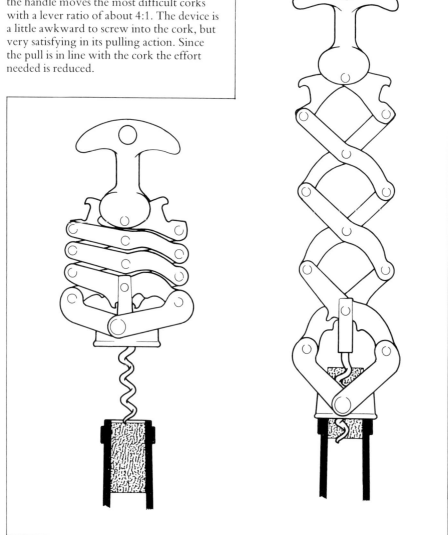

The fastest corkscrew with a lever action is a large device usually seen in bars and restaurants clamped to a bar counter. A pull on a long lever drives a nut down a screw thread rotating the corkscrew into the cork. The lever is then pushed back, this time without rotating, to extract the cork. An internal lever resets the mechanism and ejects the cork.

Screws

The corkscrew itself acts as an extractor screw on a very useful tool. An extra-long helical wire slides between two flat prongs each with a large ridge on the inside. The prongs simply flex to accommodate different bottle sizes. As the corkscrew screws into the cork the ridges bear on the tip of the bottle. A continued rotation of the corkscrew forces the cork to behave like a nut and slide upwards out of the bottle. A simple and efficient device.

7 *Some people object to the screwpull because it can deposit cork dust on the surface of the wine as the cork moves up the helix like a nut*

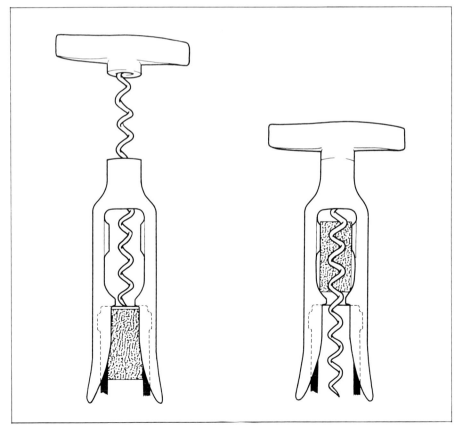

Perhaps the most popular screw system uses a left-handed screw surrounding a corkscrew. This is often made in wood or plastic. An outer case bears on the neck of the bottle and is internally threaded to carry a tube with a handle on top within which is the corkscrew with its own handle.

First the corkscrew is inserted in the usual way and then the second handle is turned in the same direction to jack the cork out. Variations on this device use a single handle to accomplish both tasks. Some use a small lever to engage the left-hand thread after the corkscrew has been inserted. Others use the friction between the bottle and rubber seat on the inner cylinder to lock them together while the thread unwinds.

8 *The double helix of viniculture.*
The large screw pulls the cork out of the bottle whilst the smaller one grips it

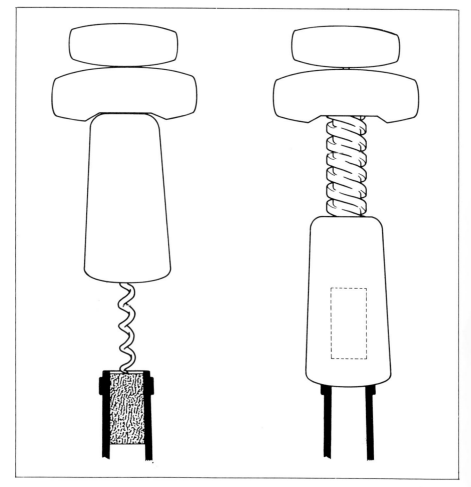

Comments

The technical content of the design of a cork extractor (or nut cracker, or knife sharpener, or other simple household tool) is apparently quite small and yet people keep finding new ways to perform the task. The judgement of their value to the user is mainly subjective, which some would think removes them from the realm of engineering. Yet it is the way in which they function that is the basic difference between the various types, and this makes them firmly a problem worthy of engineering attention. The Design Investigation on can openers (page 91) was first published in Autumn 1982 and has provoked discussion on a number of occasions on the finer points of the various types.

The fact that there are so many different functional types on the market implies that there is no one type which satisfies everyone. Indeed most people have their own favourite type which they prefer over all others. A study of these varieties is an interesting introduction to a number of mechanisms and their relative merits in a particular situation.

Design Investigation by Michael French
Professor of Engineering Design, University of Lancaster

Quarter-turn actuators

The food and process industries (and many others, too) have large, and often quite complex, pipework systems. Valves are used to control the flows through the systems, and these can perform metering, directional, or shut-off functions. Frequently some valves will have to be operated remotely, either by the push of a button on a control panel, or automatically by a computerised plant control system. The quarter-turn pneumatic actuator is often used to open and close valves in which the bore is traversed by a body, usually a ball, which can be rotated so that a hole in it is either across the flow, stopping it, or along the flow, allowing free passage (figure **1a**). The actuator is a simple engineering component which exemplifies some aspects of design very well.

Figure **1b** shows the principle of one actuator of this type, where a cylindrical casing houses two pistons, joined by a single rod carrying a rack which turns a piston on the shaft. Line air is admitted to one end to close the valve, to the other to open it.

Most designs use piston and cylinder mechanisms or vane types. This design investigation will be restricted to the piston and cylinder arrangements, which vary in the number and type of pistons and the mechanisms for converting their linear motion into a rotary one. Figure **1c** shows an arrangement having two double-acting pistons and figure **1d** shows an alternative mechanism for converting linear to rotary motion.

The choice of mechanism and configuration is closely related to the method of manufacture. One example of this relationship is the interaction between the construction of the body and the provision of air passages. A deep sheet-metal pressing, like a saucepan, will serve for a cylinder, but it cannot accommodate air passages.

Design problems can only rarely be dealt with in an orderly sequential fashion and this is no exception.

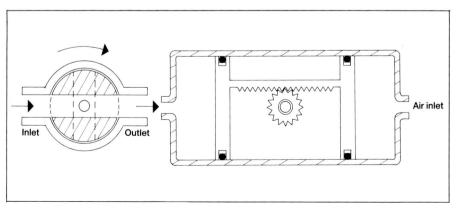

1a *Quarter-turn valve*

1b *Two single-acting pistons. Air to one end closes valve; to the other end opens it*

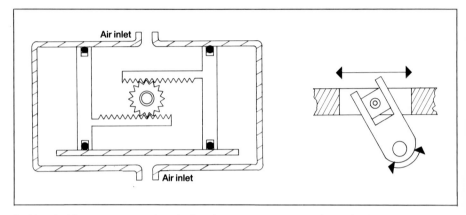

1c *Two double-acting pistons. Air to both ends to close valve (say); air to centre to open it*

1d *Another linear to rotary mechanism*

Fundamental energy considerations

If the swept volume is V_s and the gauge pressure of the supply is p, the work done is pV_s. If the torque is T and the shaft turns through $T\Theta$ (which should be a little over 90°, say 1.62 radians, to allow for tolerances) then the output work is $T\Theta$, and

$$\eta p V_s = T\Theta \qquad \text{\textit{equation 1}}$$

where η is the overall efficiency of the device, generally in the range 0.8–0.9. Thus if η is 0.8, T is 60Nm, and p is 0.6MPa (6 bar), then

$$V_s = \frac{60 \times 1.62}{0.8 \times 0.6 \times 10^6} \text{ m}^3 = 202.5\,\text{cm}^3$$

If there are N pistons working on each operation of area A and stroke s,

$$V_s = NAs \qquad \text{\textit{equation 2}}$$

Now V_{sl} as given by equation 1 with $\eta = 1$ sets an absolute lower bound to the overall volume V_o of the actuator, and so we can regard the ratio

$$\frac{V_{sl}}{V_o} = \frac{T\Theta}{pV_o} = \frac{\text{ideal minimum volume}}{\text{actual volume}}$$

$$\text{\textit{equation 3}}$$

as a measure of the merit of a particular design. The smaller the device is, the cheaper it will be; other things being equal. More usefully, we can regard this figure of merit as the product of two others, for

$$\frac{V_{sl}}{V_o} = \eta \times \frac{V_s}{V_o} \qquad \text{\textit{equation 4}}$$

The first factor, η, the efficiency, is familiar, and we are accustomed to think of high efficiency as desirable, but here its chief virtue is to make the device smaller. The second factor, V_s/V_o swept or useful volume over overall volume, is a measure of good use of space and again the advantage is smallness. As an indication, values of V_s/V_o greater than 0.25 are good: as might be expected, designs with double-acting pistons are generally better in this respect.

Linear/rotary mechanism, matching and efficiency

An important consideration in the choice of mechanism is matching. The torque required to move the valve is greatest at the ends of the travel and the mechanism of figure **1d** produces such a characteristic (figure **2**). A connecting rod and crank mechanism will produce less torque at the ends of the travel and this is probably a good enough reason to reject it. The rack and pinion mechanism gives a constant torque, neglecting a small variation in efficiency. Thus on matching alone we would choose figure **1d** in which the maximum torque is $\pi/2$ times the average, neglecting friction.

However, the introduction of friction reduces the relative advantage over the rack-and-pinion and the construction is awkward and expensive, with high stresses in intricate components. Efficiency at the ends of the stroke, where it counts, is likely to be about 74% with a coefficient of friction μ of 0.15; whereas the rack and pinion may be 85% efficient.

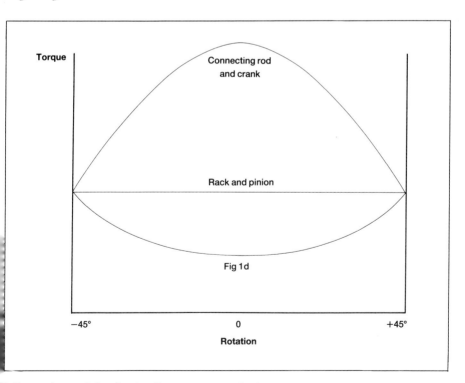

2 *Torque characteristics of various linear to rotary mechanisms*

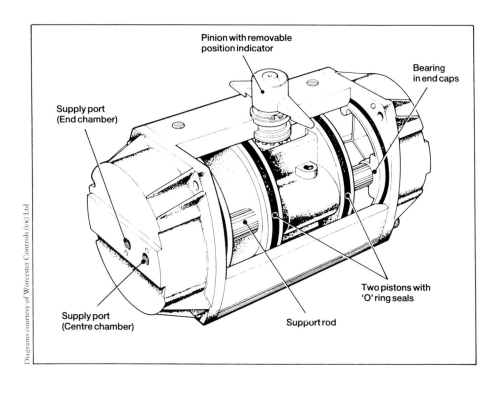

Pinion with removable
position indicator

Bearing
in end caps

Supply port
(End chamber)

Two pistons with
'O' ring seals

Supply port
(Centre chamber)

Support rod

Operational Sequence

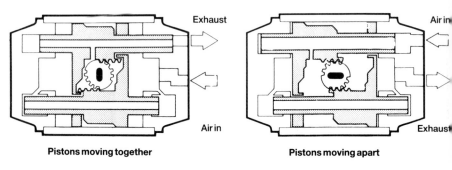

Exhaust

Air in

Air in

Exhaust

Pistons moving together

Pistons moving apart

Worcester Controls double-acting pneumatic actuator

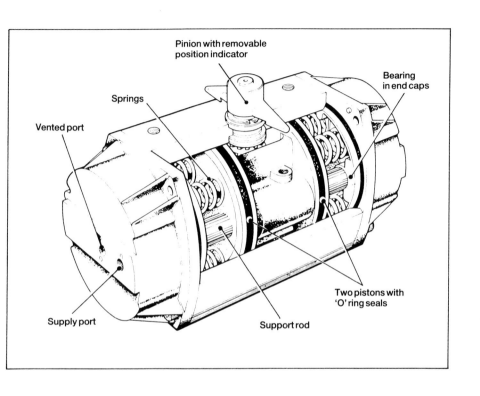

- Pinion with removable position indicator
- Bearing in end caps
- Springs
- Vented port
- Supply port
- Support rod
- Two pistons with 'O' ring seals

Operational Sequence

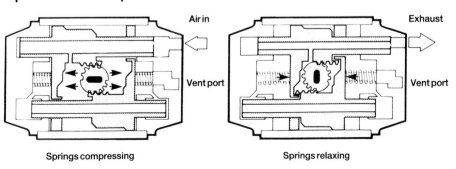

Air in

Vent port

Springs compressing

Exhaust

Vent port

Springs relaxing

Worcester Controls two single-acting piston pneumatic actuator with spring return mechanism

Calculation of efficiency

This is not easy, so figure **3** shows how to do it graphically (this is one of those cases where graphics may still be best). The air force **F** on the piston tends to twist it, producing reactions at **R** and **S**. Notice the extension at **R**, running on the cylinder wall, which provides the improved guidance that is essential to make figure **1c** work. The reactions at **R** and **S** are inclined to the left at the angle of friction λ, tending to oppose the rightward motion of the piston. **P** is the reaction from the pinion teeth at the pressure angle ϕ (a more detailed treatment would include friction at the mesh). The piston-cum-rack is in equilibrium under the four forces **F** and **P** (whose lines of action intersect at **N**) and **R** and **S** (whose lines of action intersect at **M**). **R** and **S** have a resultant through **M**, and the body is in equilibrium under the three forces **F**, **P** and this resultant, which must pass through **N**. Hence we can draw a triangle of forces for **F**,**P** and the resultant (**XVW**, figure **3**). If λ = 0, the efficiency is 1, **R** and **S** are parallel so **M** is at infinity and **MN** is perpendicular to the axis: the triangle of forces becomes **XUW**, with the turning force **P** increased from **VW** to **UW**. Hence

$$\eta = \frac{\mathbf{VW}}{\mathbf{UW}}$$

and this can be about 0.9 with a long guidance length and low values of μ.

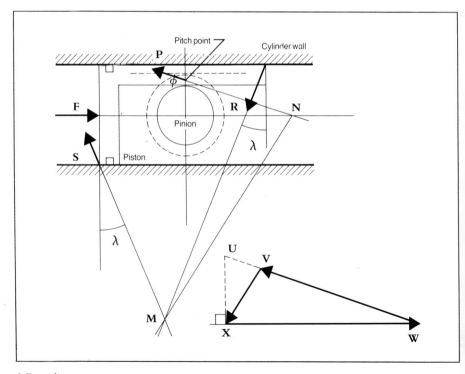

3 *Force diagram*

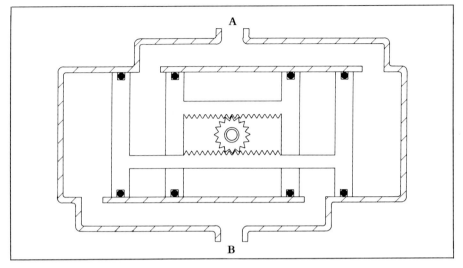

4 *Air to port A turns shaft clockwise; air to port B turns shaft anticlockwise. This curious scheme has good V_s/V_o for one with an unpressurised shaft*

Long guidance lengths give stacking problems with increased device length and hence reduced figures of merit. Low friction coefficients are associated with materials of low strength or lower allowable surface loading, which brings us to look at materials and method of manufacture.

Materials and manufacture

The choice of materials is dictated by factors such as cost, strength, corrosion, manufacture etc. In practice the choice is limited to steel, stainless steel, aluminium alloys, polymers (plain or reinforced) and a few others for details like seals and perhaps bearings. Most materials come in many forms and the final choice, for example of a polymer, takes much study. Here only a few simple considerations will be outlined.

Suppose the cylinders are to be made in aluminium alloy. Extrusion has its attractions: the metal is of high quality with no porosity and the external mounting faces do not require machining. An extrusion is often cheaper than a casting. On the other hand with a casting it is possible to provide local bosses for the shaft bearings or air connections. Such bosses may be needed where the loads on the pinion are unbalanced, for example, in devices with single-acting pistons.

Now consider the pistons. It would be attractive to make these in a polymer, but most polymers are not strong enough for the duty of the rack teeth. If the piston and rack are to be in one material, it probably has to be an aluminium alloy. We could use a polymer for the piston with an aluminium alloy insert for the rack; on the other hand for the insert we might use some other material more suitable for the job than aluminium alloy.

Coefficients of friction are important, and here polymers or polymer coatings working against metals or 'foreign' polymers are desirable.

(F) Function or Aspect	Options	Comments
1 Configuration	1.1 One double-acting piston	1.1 bulky, simple, leads to trouble with F4
	1.2 Two single-acting pistons	1.2 fairly simple, moderate bulk
	1.3 Two double-acting pistons	1.3 complex but small; no journal loads on shaft
2 Contain air	2.1 Aluminium alloy extrusion	2.1 cheap, but not suited to 1.1 or 1.2
	2.2 Aluminium alloy casting	2.2 fairly cheap
	2.3 Stainless steel pressing	2.3 cheap, if F4 can be arranged cheaply
4 Distribute air	4.1 Passages in walls	
	4.2 Passages in piston rods	
	4.3 Hollow guide rods	4.3 good where guide rods available already
	4.4 External pipes	4.4 not acceptable to market
	4.5 Hollow tiebolts	
6 Guide pistons	6.1 Cylinder walls	6.1 usual solution
	6.2 Guide rods	6.2 good, but expensive if not turned to advantage; not much of a problem with 2.2
7 Secure cylinders or cylinder ends axially	7.1 Screws	7.1 usual solution
	7.2 Ring nut	7.2 bulky, expensive; awkward
	7.3 Circlip	
	7.4 Split central casing	7.4 not suitable where centre pressurised (2.3)
	7.5 Tie bolts	7.5 see 4.5

Table of options

Normally an analysis of functions and a table of options should be made earlier on in the problem, but for the present purpose this seemed the right place to introduce it. Many of the functions and options have been left out, partly for brevity but also because this is a kernel table of the kind to which the designer should prune his original, unmanageable table as his insight develops. The left-hand column also contains configuration, which is not a function but must be included.

For example, suppose a two single-acting cylinder configuration is chosen, 1.2 in the table, as in figure **1b**. Because the centre space is not under pressure it is feasible to use a split centre casing embracing two stainless steel pressed cans: a very economical and pleasing solution to 2 and 7. However, there are then difficulties with function 4,

where only 4.2 is available. Worcester Controls use this scheme in their Miti-1 actuator, transferring air from the central casing to passages in the piston rod via holes and slots bordered by O ring seals. The pistons-cum-rod component is in acetal. with a sintered stainless steel insert carrying the rack teeth.

A much older Worcester Controls design, still a market leader today, uses two light alloy double-acting pistons rigidly fixed to stainless steel guide tubes. These guide tubes slide in polymer bushes in the aluminium alloy end caps, so that the only metal-to-metal contact with the pinion occurs at the racks. Because the loads on the pinion form a pure couple, there are no journal loads on the pinion shaft and this makes it practical to make the body from an extrusion; the large journal loads in the single-acting two piston configuration require long bearings (bushes) to carry them.

Producing good designs

One good approach is to make a preliminary decision, for example choose the two double-acting cylinders: they are compact, with balanced loads on the pinion. Then go on to seek happy combinations of choices on the other functions, for instance ask yourself what advantages there might be in choosing an aluminium casting for the body, rather than the extrusion option; one designer found a good chap way of fixing the end caps. Remember there are usually only a few choices in design which can be made independently of other decisions. Remember also that the good design has many virtues and no real vices.

Finally here are a couple of ideas you may like to look at. One complication (there are many) which has been omitted for the sake of brevity is that of spring returns. Many valves are required to return to a particular position, on or off, in the event of failure of air supply pressure, and this is often achieved by fitting compression coil springs acting on the appropriate piston or pistons. Now if the minimum torque required is T, then the springs must be strong enough to produce this at the end of their stroke. Because the force in the springs is higher when they are most compressed, the torque they produce at the beginning of the stroke will be about 2T. Finally the air pressure on the other side of the piston or pistons must be capable of overcoming this 2T of torque and providing a full additional T for the valve, ie when spring return is needed the actuator must be roughly three times as large for the same torque. Also there is no need to apply air pressure to the spring-loaded side of the piston or pistons.

It is worth considering an actuator that always uses a spring return. It should be cheaper than a normal spring return actuator, since it would only have one air side. In competition with the usual actuators without spring return it would probably be more expensive, being about three times bigger, and the saving due to reduced complexity would hardly be enough to compensate, even neglecting the cost of the springs themselves. It is probably not a winner, but the idea is interesting. Do not forget that it must be possible to make the spring-operated actuator turn either way, eg, by turning it upside down.

The other idea is the curious scheme in figure **4**, where each of the racks/piston rods has a piston at each end. The outer pair of pistons are double-acting and the inner pair are single-acting. A complication occurs where the piston rod joining the outer pair of pistons goes through the inner pair of pistons, but with ingenuity it can be overcome. The rack forces on the pinion do not balance, but the bearing loads are only a third of those in the scheme of figure **1b**. The arrangement is complicated in function, but has some advantages.

Bibliographies

Bibliography 1: books

Finding books on engineering design is often difficult. Even books actually called Engineering Design *often turn out to be textbooks on structural analysis or a similar field with hardly a mention of genuine design. As a result the compilation of a bibliography was included as part of this teaching aids programme. It consists of works in English extracted, with permission, from the* International Bibliography of Design Science, *which is edited by Dr V Hubka of Zurich and published by Heurista.*

A

ALGER, J R and HAYS, C V, *Creative Synthesis in Design*. Englewood Cliffs, Prentice-Hall, 1964.
Engineering and design (nature of design, design process). Design process (steps, design questions). Creative idea origination, scheduling innovation, envisioneering the future.

ALLAN, J J, *CAD Systems*. North-Holland, 1977.
Proceedings of the IFIP Conference 1976. CAD systems, methods, techniques. Artificial intelligence in CAD. The design of CAD systems. Data structure. Interactive graphics.

AQUILAR, R J, *Systems Analysis and Design in Engineering Architecture, Construction and Planning*. Englewood Cliffs, Prentice-Hall, 1973.
Systems-approach. Economic considerations. Deterministic systems (linear, integer and dynamic programming). Stochastic systems (linear, dynamic programming, theory of games).

ARCHER, L B, *Design Awareness and Planned Creativity in Industry*/Connaissance du design et de la creativite planifiee dans l'industrie. London, Design Council, 1974.
Some definitions. What is 'good design?' Design and conflict of interest. Design as an element in corporate strategy. Industry and industrial designer.

ARCHER, L B, *Technological Innovation – a Methodology*. SPF – special publication series. London, Inforlink, 1971.
On the nature of value. On the nature of problems. A generalised model for problem statements. An algorithm for problem solving. A characteristic project program. Definitions.

ARCHER, L B, *The Structure of Design Process*. London, Royal College of Art, 1969.
Definition of design. Systematic model. Operational model. Logic of design procedure. Problem of aesthetics. Techniques in problem solving.

ASHFORD, F, *The Aesthetics of Engineering Design*. London, Business Books, 1969.
The function of aesthetics in engineering design: perception, form, aesthetic manoeuvre, surface treatment. Glossary.

B

BAKEY, T, *Economic Benefits and Economic Impact of Interactive Computer-aided Design*. American Society of Mechanical Engineers Publication 79–PVP–80, 1979.
System approach – manual versus CAD. Average productivity comparisons. Turnaround time. Error proneness. Office space. Operating costs. Pay-off period.

BARTEE, E M, *Engineering Experimental Design Fundamentals*. Englewood Cliffs, Prentice-Hall, 1968.
The methodology of design. The analysis phase. The evaluation phase.

BLACK, M, *Engineering and Industrial Design*. Proceedings, Institution of Mechanical Engineers, vol 186, no 74, 1972, p 72.
A confusion of terms. Style and styling. The morality of industrial design. Education for design. The future of industrial design.

BLUME, P, *Computer Aided Design*. Philips Technical Review, 1976.
Stages in design process. Workpiece – descriptive language. Input via display. Geometrical structure. Data structure. Integrated CAD/CAM system. Economic aspects.

BRAITHWAITE, W, *Boeing's CAD/CAM Integrated Information Network*. New York, American Institute of Aeronautics and Astronautics, 79–1847.
Computer equipment. Definition of surfaces. Intersections. CAD/CAM environment. Data communication and management.

BRICHTA, A M and SHARP, P E M, *From Project to Production*. Oxford, Pergamon, 1970.
Concept of development. Rationalisation of project work. Engineering design. Optimum

design. Inventions, patents and design registration. Design realisation. Material and stress analysis. Models and prototype. Launching a product.

BRILHART, J K, *Effective Group Discussion.* Dubugne, Brown, 1973.
Discussion. Orientation to groups. Organising group discussions. Preparing to discuss. Discussion leadership. Participating. Communication. Observing and evaluating discussions.

BUBENKO, J, LANGEFORS, B and SOLOBERG, A, *Computer-aided Information Systems Analysis Design.* Copenhagen, Lund, 1972.
Project conception. System development. Project NO. Political decisions in design of information. Processing systems. Datamatics/informatics. DIFO: design of information systems. Cascade computer-based documentation system. CADIS: computer-aided design of information systems. Development. Computer-graphics as tool.

BUCK, C H, *Problems of Product Design and Development.* London, Pergamon, 1963.
Initiation. Idea. How will we sell? Function and use. Design for production. Distribution. Maintenance. Co-ordination. The designer. The drawing office. Research. Legal protection.

C

CAIN, W D, *Engineering Product Design.* London, Business Books, 1969.
Definitions. Type of product. Skills and requirements. Factors. Investigation. Concept. Function, use, appearance, production. Standardisation. Specifications. Tests. Value of materials. Manufacturing process, methods.

CHESTNUT, H, *Systems Engineering Methods.* New York, Wiley, 1967.
The environment for systems engineering methods. System organisation. Scheduling and record-keeping. Formulating and structuring the system. Factors for judging value of a system. Cost. Time. Reliability. Conclusion and prologue.

CHIRONIS, N P, *Mechanisms, Linkages and Mechanical Control.* New York, McGraw-Hill, 1965.
Compilation of mechanical, electrical, hydraulic, pneumatic, optical, thermal and photoelectric devices for a wide variety of functions.

CHOW, W, *Cost Reduction in Product Design.* New York, Van Nostrand, 1978.
Engineering concepts of efficient design. Mathematical methods. Economic factors. Design for strength with minimum material. Simplifying assembly. Manufacturing methods. From theory to praxis.

CLAUSEN, H, *Engineering Research, Development and Design.* London, Clausen, 1965.
Science museum revisited. Training for design.

Fruits of past changes in the structure of engineering education. R and D. Professional engineers and technicians. Engineers and draughtsmen. Management and leadership.

CREAMER, R H, *Machine Design.* 2nd ed. London, Addison-Wesley, 1976.
Mechanics and strength. Friction. Bearings. Shafts. Fasteners and welding. Belting. Chain drive. Brakes. Clutches. Gears. Springs. Power units. Appendices.

CRICKMAY, C L and CHRISTOPHER J, *Design Imagination and Method. Designing as a Response to Life as a Whole.* Open University Press, 1972.
Designing in a wider context. Skills for designing in a wider context. Skills for acting widely. Example of exploring problem structure. Meta cards.

D

DIXON, J R, *Design Engineering. Inventiveness, Analysis and Decision Making.* New York, McGraw-Hill, 1966.
Inventiveness. Engineering analysis. Decision making. A hundred problems, case studies. Twenty-seven project proposals, problem formulation, physical principles, checking, computation, evaluation. Generalisation.

DOBROVOLSKY, V, and others, *Machine Elements.*
Methods, practice, machine elements, choice of materials, degree of accuracy machining, various types of joints, assemblies, transmissions.

DUSSEILLER, B, *ICEPS 1 – a Procedure for Evaluation of Innovations. Method and Experience.* Sulzer Technical Review 61, 1979.
The ICEPS method (development stages, procedure steps). Evaluation procedure. Practical experience with ICEPS.

E

EASTMAN, C H M, *Exploration of the Cognitive Processes in Design.* Pittsburgh, Carnegie Mellon University, 1968.
Introduction. A background study. Experimental research in design methodology. Operational description of design process. Original notes of designer. Activity graph.

EDEL, D H, *Introduction to Creative Design.* Englewood Cliffs, Prentice-Hall, 1967.
Basic concepts. Factors influencing. Material and human factors. Related procedures. Creative thinking factors. Synthesis. Invention, patents. Design process. Detailed design for production. Case study.

EDER, W E and GOSLING W, *Mechanical System Design.* Oxford, Pergamon, 1965.
Systems. The design process. The system

breakdown. Consideration of detail. Strength calculations. The human element. Completing the design.

EDHOLM, O G, *The Biology of Work*. London, World University Library, 1967.
Muscular work. Work and body heat. Climate. Energy and food. Sight and vision. Sound, noise, vibration. Display of information. Layout of equipment. Training. Shift work. Fatigue. Accidents. Age at work. Bibliography.

ELLINGER, J H, *Design Synthesis*. Vol 1 and 2. London, Wiley, 1967.
Vol 1: Development by iteration. Design aids. Design example. Conclusion.
Vol 2: Design of a carriage for shiptank.

F

FEILDEN, G B R, *Engineering Design-Report*. London, Her Majesty's Stationery Office, 1963.
Definitions. The importance of engineering design in the national economy. The present standing of engineering design. Factors affecting the quality of engineering design. The remedies.

FLURSCHEIM, C H, *Engineering Design Interfaces and Management Philosophy*. London, Design Council, 1977.
Interfaces in engineering design. Commercial engineering. Interfaces within technical functions. Design standardisation interface. Design quality. Design innovation. Man/machine interface.

FOX, R L, *Optimisation Methods for Engineering Design*. Reading (Mass), Addison-Wesley, 1971.
Introduction to formulation of optimisation problems. Unconstrained minimisation. Constrained problems by unconstrained minimisation. Direct methods formulation. Constrained problems special techniques and applications.

FRENCH, M J, *Engineering Design: the Conceptual Stage*. London, Heinemann, 1971.

G

GAGNE, R M (editor), *Psychological Principles in System Development*. New York, Holt, Rinehart and Winston, 1962.
Fourteen articles or chapters on man and system, tasks, job, training, teams.

GEIGER, G H, *Supplementary Readings in Engineering Design*. New York, McGraw-Hill, 1975.
Engineering construction contracts. Patent law. Safety standards. Codes and practice for plant design. Industrial risks. Designing plants to meet OSHA standards. Efficient effective writing.

GILBERT, A R, *Designing for Form as well as Function*. American Society of Mechanical

Engineers Technical Paper 73–DE–6, 1973.
Fundamental aesthetic design considerations.

GLEGG, G L, *The Design of Design*. Cambridge University Press, 1969.
The design of the problem. The designer. The design of design: the inventive. The design of design: the artistic. The design of design: the rational. Safety margins.

GLEGG, G L, *The Science of Design*. Cambridge University Press, 1973.
Carrying out the scientific research needed to obtain data for engineering design.

GREENWOOD, D, *Mechanical Details for Product Design*. New York, McGraw-Hill, 1961.
Tables and charts. Accessories. Basic and general design. Control and materials handling. Fastening and joining. Hydraulics and pneumatics. Mechanical movements and linkages. Mechanical power transmission. Spring devices. Welding and brazing.

GREGORY, S A, *Creativity in Chemical Engineering Research*. Proceedings, Symposium on Productivity in Research, London, 1963.
Creativity and the environment. Outline of creative activity. Creative person. Education and creativity of individuals. Creativity in groups. Heuretic methods. Problem solving.

GREGORY, S A, *Graphical Interactive Language for the Engineering Draughtsman and Manufacturing Engineer*. National Engineering Laboratory Glasgow Paper 2, 7, June 1978.
GRIP-language developed by Unit Computing Corporation, subsidiary of McDonnell Douglas Automations (MCAUTO). Computers used in interactive CAD/CAM systems to perform scientific calculations.

GREGORY, S A and BRIDGWATER, A V, *Some Approaches to New Process Design Concept Generation and Evaluation*. Journal of Chemical Education Symposium Series no 35, 1972.
Morphology applied to process design.

GREGORY, S A, *The Design Method*. London, Butterworth, 1966.
Design method in practice. Definitions and methodologies. Human perspective. Ergonomics and design. Elements of design. Design techniques. Management and design. Design and research.

H

HAJEK, V G, *Product Engineering (Profitable Technical Program Management)*. New York, McGraw-Hill, 1965.
Specification. Selection of approach. Contract. Writing technical proposal. PERT. Costs. Negotiations. Project initiating, monitoring. Engineering design. Reliability. Test.

HANSEN, F, *Comments on Scientific Designing*

in Seminarium Nauki Konstrukcji. Gliwice Politechnika, 1972.

HARRISBERGER, L, *Engineersmanship (A Philosophy of Design).* Belmont, Wodsworth, 1968.
Design horizons. Design attitude. Creativity. Design process. Value engineering. Style and aesthetics. Communications and selling.

HILL, P H, *The Science of Engineering Design.* New York, Holt, Rinehart and Winston, 1970.
New products and the corporative environment. The creative process. Design process. Three case stories. Materials selection. Human factors CPM/PERT. Value engineering. Patents, critique, projects.

HOLT, K, *Product Innovation – Models and Methods.* The Section of Industrial Management, The Norwegian Institute of Technology, Trondheim, R 17, 1975.
Part I: Terminology and models. Generation of ideas. Utilisation of ideas. Preparation for implementation. Manufacturing and marketing.
Part II: 95 tools for product innovation.
Part III: 10 papers on product innovation.

I

IRONS, B and AHMAD, S, *Technique of Finite Elements.* New York, Wiley, 1980.
Basic techniques. Organisation techniques. Trends in element formulation and in solution techniques. Speculations. Theoretical details.

ISAAC, A R G, *Approach to Architectural Design.* London, Iliffe Books, 1971.
Basic perception (Shape, symmetry and balance. Movement and attention, continuity, colour.) Design principles (design process, planning sequence).

J

JERGER, J J, *Systems Preliminary Design.* Princeton, Van Nostrand, Kibo.
Reliability of missile systems. Guidance and control design data. Kinematics. Homing missiles. Heat transfer. Stability. Structures design procedure.

JOHNSON, R C, *Mechanical Design Synthesis.* New York, Van Nostrand Reinhold, 1971.
Part I: Synthesis of configurations: creative design, aids in creative effort, selection of optimum configuration.
Part II: Selection of materials and dimensions: shape design, advance design, advanced design of elements and systems (12 examples).

JONES, J C, *Conference on Design Methods.* London, Pergamon, 1963.
Relevance of system engineering. Methodology for design of instruments. Design method in architectural education. Method of system design.

Problems of design of a design system. Creative process and methods. Psychological aspects of the creative act.

JONES, J C, *Design Methods – Seeds of Human Futures.* London, Wiley-Interscience, 1970.
The developing design process. Design methods in action. Methods of exploring design situations, methods of searching for ideas, methods of exploring problem structure, methods of evaluation.

JONES, S W, *Engineering Design Problems.* London, Iliffe Books, 1969.
Seventeen problems: eg Pulley. Journal bearings. Shaft. Drive with geared reduction. Gearbox. Valve. Lifting device. Bolted and welding joints.

JORDAN, S (editor), *Handbook of Technical Writing Practices.* Vol 1 and 2. New York, Wiley, 1971.
Military and commercial applications. Governmental, industrial and consumer coverage. Heavy emphasis given to combined aerospace and electronic industries and a wide spectrum of technological disciplines, industrial groupings etc.

K

KASSOUF, S, *Normative Decision Making.* Englewood Cliffs, Prentice-Hall, 1970.
Decision making under certainty: with objective probabilities; with subjective probabilities. An application portfolio selection. In the absence of probabilities. Further generalisations.

KIRK, F, *Total System Development for Information Systems.* New York, Wiley, 1973.
System development, factors, need, life, purpose, use, project team. Activity network, eight phases. Data collection techniques, human factors, standards, training, technology.

KOLL, *Constronic 76, 2nd Conference on Mechanical Aspects of Electronic Design, September 1976.* Budapest, DMKDK-Technoinform, 1976.
Education effects of new results in technology, CAD. Thermal problems. Reliability, environmental stability. New alternatives in mechanical design. Problems of mechanical engineering. Standardisation, typifying.

KOLL, *Constronic 80, 3rd Conference on Mechanical Aspects of Electronic Design, April 1980.* Budapest, DMKDK-Technoinform, 1980.
A, Methodology and effectiveness of design. B, CAD. C, New technology. D, Thermal problems. E, Standardisation. F, New alternatives in mechanical design. G, Education.

KOLL, *Information Systems for Designers 1971.* Symposium July 1971, Southampton. Southampton, The University, 1971.
Fifteen papers from: Holmstrom, E; Shaw, J A;

Vickers, J S; Harries, M V; Wall, R A; Benson, J P; Benning, V J; Slade, I M; Bottle, R T; Tell, V B; Jarvis, A S; Cosier, P H; Pitts, G; Cousins, L B; Billheimer, J S.

KOLL, *Information Systems for Designers 1974.* Meeting July 1974, Southampton. Southampton, The University, 1974.
Sixteen papers from: Grant, D P; Ware, J E; Patrick, A F; Raman, P G; Nordstrom K; Ray-Jones A; Pitts, G; Ouye, J A; Wall, R A; Robinson, F; Leather, G M; Pugh, S; Webster, J; Hulme, G; Hills, P; Dehlinger, H.

KOLL, *Information Systems for Designers 1977.* 3rd Symposium March 1977, Southampton. Southampton, The University, 1977.
Twenty-eight papers from: Highson, J; Davies, T H; Lee, J H; Potier, O N R; Harries, D B; Owen, B S; Martin, F A; Pugh, S; Pitts, G; Brewer, R; Turner, B T; Hills, P C; Fielding, J P; Smart, B L; Waller, G; Powell, J A.

KOLL, *Information for Designers 1979.* 4th Symposium July 1979, Southampton. Southampton, The University, 1979.
Fourteen papers from: Druce, G; Fricker, D; O'Kelly, J; Bottle, R T; White, M S; Riemer, G K; Mahmoud, D; Pugh, S; Hills, P C; Fisher, B C; Wingfield, L; Hodson-Smith, C; Smart, B L; Pitts, G; Moss, T R.

KOLL, *The Design Method.* Symposium September 1965. London, Butterworth, 1966.
Papers on methods (3p), human perspective (13p), techniques (10p), management and design (5p), design research (4p).

KOLL, *The Theory of Machines and Mechanisms.* Proceedings of the 5th Congress, July 1979, Montreal. New York, ASME, 1979.
Kinematic analysis and synthesis (25p), technology transfer (5p), plant engineering (6p), CAD (7p), dynamics of gears (7p), rotor dynamics (10p), isolators and dampers (8p), problems in turbomachinery, etc.

KRICK, E V, *The Introduction to Engineering and Engineering Design.* New York, Wiley, 1969.
Objectives. Qualities of the competent engineer. The design process. The design cycle.

L

LATHAM, R L, *Problem Analysis by Logical Approach System (PABLA).* Aldermaston, AWRE, 1975.
Introduction. What is PABLA? Definition of design. How PABLA works. The system in detail. Origin and history of the system. Synopsis of a logical design procedure.

LAYTON, C, *Ten Innovations.* London, Allen and Unwin, 1972.

Managing innovation and the use of qualified scientists and engineers. Ten industries: the case studies.

LEECH, D J, *Management of Engineering Design.* New York, Wiley, 1972.
Defining the problem. Feasible solutions. Making the hardware. The management of design. Models and tools of analysis.

LIFSON, M W, *Application of Criteria and Measures of Value in Engineering Design.* Dissertation, University of California, Los Angeles, 1965.
Model of the design process. Decision and utility theory. A design value model.

LITTLE, J F, *Critical Thinking and Decision Making.* Toronto, Butterworths, 1980.
Meaning: extracting and clarifying claims. Structure: exhibiting and testing scheme. Justification: evaluating evidences. Decision making and articulation.

LUZANDER, W J, *Basic Graphics: for Design Analysis, Communication and the Computer.* New York, Prentice-Hall, 1968.
Basic graphical techniques. Spatial graphics. Graphics for design and communication. CAD and automated draughting.

M

MARKS, D H, *System Planning and Design. Case Studies in Modelling, Optimisation and Evaluation.* Englewood Cliffs, Prentice-Hall, 1974.
Introduction. Systems modelling. Optimisation. Evaluation.

MATCHETT, E, *The Accelerated Development of Creative Mental Skills.* Industrial Training International, 1968.
Fundamental design methods. Development of a less skilled person. Formation of design skill development group.

MAYALL, W H, *Industrial Design for Engineers.* London, Iliffe Books, 1967.
Ergonomics and industrial design. Visual effects of line and form. Colour. Aesthetic concepts. Style. Industrial design in practice.

MAYALL, W H, *Principles in Design.* London, Design Council, 1979.
The makings of design. The principles of totality, time, value, resources, synthesis, iteration, change, relationships, competence, service. Glossary.

MEREDITH, D D and others, *Design and Planning of Engineering Systems.* Prentice-Hall, 1973.
System approach. Modelling and analysis. Optimisation. Programming. Organisation. Networks. Decision analysis systems simulation. Planning. Project management systems theory.

MERRYWEATHER, H, *Two Application Programs which Link Design and Manufacture.*

National Computer Conference, 1975.
GNC–2 and 2½D package for verifying design.
Production of numerical control tapes and
checking manufactured components.
POLYSURF–3D package for designing, drawing
and producing of numerical control tapes. Both
packages developed by CAD Centre at Cambridge.
Development, facilities and examples.

MESAROVIC, M D, MACKO, D and TAKAWA, Y,
Theory of Hierarchical Multilevel Systems.
Academic Press, 1970.
Examples. Conceptionalisation. Formalisation
and coordination. Mathematical theory of
coordination. Unconstrained optimising systems.
Optimal coordination of dynamic systems.

MIDDELTON, M, *Group Practice in Design.*
New York, Braziller, 1967.
The nature of design. Patterns of collaboration.
Communication. Product design.

MILLER, D W and STARR, M K, *The Structure
of Human Decisions.* Englewood Cliffs,
Prentice-Hall, 1967.
Science and administration. Responsibility for
decisions. Objectives of decisions. Structure and
analysis of decisions. Applied decision theory.
When is a problem worth solving?

MISCHKE, C R, *An Introduction to Computer-
aided Design.* New York, Prentice-Hall, 1968.
Engineering design and the digit. Computer.
Figures of merit. Computer requirements. The
search for extremes. Problems as they confront
the engineer. Appendix: problems.
Documentation of IOWA-CADET subroutines.

MORRISON, D, *Engineering Design.*
New York, McGraw-Hill, 1968.
The nature of engineering design. Designs
obtainable by a series of independent systems.
Practical design as an approximation to ideal
design. Bearings. Design of inherently stable
devices. Application of elementary information
theory to mechanical design.

MUDGE, E, *Value Engineering.* New York,
McGraw-Hill, 1971.
Fundamentals and theory. Basic terms. Overview
of the system approach. General phase, information
phase, function phase, creative phase. Evaluation.
Investigation. Recommendation. Project selection.
Case studies (25).

MUELLER, E, *Development of a Mechanical
Design Using Basic Elements.* Guildford, IPC
Business, 1976.
Problem arising during development of a
construction algorithm. Construction using basic
elements. Adaptation of structure and technically
orientated structural adaptation.

MURREL, K F H, *Ergonomics.* Chapman and
Hall, 1969.
Human body. Design factors. Environmental
factors. Organisational factors.

N

NEUFVILLE, R (editor), *Systems Planning and
Design.* Englewood Cliffs, Prentice-Hall,
1974.
Case studies in modelling, optimisation and
evaluation. Systems modelling. Optimisation.
Evaluation.

NEWMAN, W M, *Trends in Graphic Display
Design.* Institute of Electrical and Electronic
Engineers Transactions on Computers 25,
1976.
Development of line-drawing displays. Picture
definition. Transformation. Display generation.
Raster-scan displays. New display devices.

NEWMAN, W H and SPROULL, R F, *Principles
of Interactive Computer Graphics.*
McGraw-Hill, 1973.
Computer graphics. Input-output equipment.
Information display systems. Programming.
Examples. Exercises.

P

PAPANEK, V, *Design for the Real World.*
London, Thames and Hudson, 1972.
What is design? Social and moral responsibilities
of the designer. Creativity versus conformity. Use
of biological prototypes in design of man-made
systems. Design and environment. Education of
designers and the construction of integrated
design teams.

PARE, A L, *A Computer Algorithm to Design
Compound Gear Trains for Arbitrary Ratio.*
ASME, 1971.
Introduction. Brief review of available methods.
The proposed computer algorithm. Numerical
example. Extension to the triple-pair gear train
problem.

PARE, E and others, *Introduction to Engineering
Design.* New York, Holt, Rinehart and
Winston, 1963.
1 Facts of engineering design. 2 Graphic
representation. 3 Orthographic projection,
pictorial drawings. Auxiliary views. Sectional
views. Size specification. Machine element
materials. Production, component design.
Kinematics. Design application.

PARK, W R, *Cost Engineering Analysis.* A guide
to the economic evaluation of long projects.
New York, Wiley, 1973.
Value of money. Investment. Cash-flow analysis.
Depreciation. Cost of capital. Cost analysis. Cost
estimation. Equivalent annual cost. Break even
and financial analysis. Forecasting. Probability.
Risk. Economic models.

PARNES, S J, *Student Workbook for Creative
Problem-Solving Courses and Institutes.* New
York, State University, 1963.
Introduction of student. Worksheets. Example
of the creative problem. Solving process.

Application to a problem. Illustration of application of idea-spurring questions to a problem.

PARR, R E, *Principles of Mechanical Design*. New York, McGraw-Hill, 1970.
Design process. Materials in design. Manufacture process in design. Design considerations and procedures. Stress analysis. Dimensions. Tolerances and fits. Design examples and projects.

PEARCE, P, *Structure in Nature is a Strategy for Design*. London, MIT, 1978.
Introduction. Structure in nature and design. Principles of built structures. Ordering principle and geometry. Theory of special order. Theory of structure. Reduction to practice. Environmental structure.

PEAT, A P, *Cost Reduction Charts for Designers and Production Engineers*. Brighton, Machinery Publishing Company, 1968.
General principles. Use of the charts. List of elements. Complexity chart. Appendix of eight elements.

PHELAN, R, *Fundamentals of Mechanical Design*. New York, McGraw-Hill, 1962.
Motion in machines. Dimension determination. Mechanical fasteners, springs, couplings, clutches, brakes. Belt and rope drives. Chain drives. Mechanisms. Bearings, Appendix.

PILDITCH, J and SCOTT, D, *The Business of Product Design*. London, Business Publications, 1965.
Organisation, planning, research, economics, maintenance, costs, marketing, new materials and processes.

PILKEY, W D, *Optimum Mechanical Design Synthesis*. Washington, US Department of Commerce, 1966.
Introduction. Technical discussion. Conclusions. Recommendation. Appendix (absolute optimum design configuration. Optimisation of parameters in a complex design. Configuration.)

PITTS, G, *Techniques in Engineering Design*. London, Butterworth, 1973.
The design process. Economic implications of design. Digital computer as a design aid. Numerical description of curves and surfaces. Some aspects of optimisation.

PYE, D, *The Nature of Design*. London, Studio Vista, 1969.
Art and science. Invention and design distinguished. Six requirements: geometry, techniques, improvement, economy, compromise, workmanship. Architecture. Inventing the objects. 'Function' and fiction.

R

ROADSTRUM, W E, *Excellence in Engineering*. New York, Wiley, 1967.
What engineering is. The engineer. Project and project team. Project control. Drawing and reports. Problem solving. Laboratory work. Design: manufacturing research and development systems. Creativity.

ROBERTS, D I, *Detailing a New Diesel Engine Design using a 3D-Interactive Graphic System*. ASME, July 1976.
Generation of true 3D-database. Producing drawings, numerical control machining tapes and finite-element models by Detroit Diesel, Allison Division.

ROE, P H, SOULIS, G W and HANDA, V K, *The Discipline of Design*. Boston, Allyn and Bacon, 1967.
Concept of planning and of system. Innovation and creativity. Role of information in design. Decisions. Value and utility. Morphology. Definition of design problem. Generation of alternatives. Feasibility. Models and simulation. Analysis. Optimisation.

ROSENAUER, N and WILLIS, A H, *Kinematics of Mechanisms*. New York, Parr, 1967.
Kinematic pairs and chains. Euler-Savay. Bobillier. Velocity. Acceleration. Relative motion. Velocity an acceleration construction. Synthesis of mechanisms.

ROSENSTEIN, A B, RATHBONE, R R and SCHNEERER W F, *Engineering Communications*. Englewood Cliffs, Prentice-Hall, 1964.
Communication: role, mathematical theory. Communication systems. Communication practice, reader, writer, report, oral reporting, graphics in engineering design. How to sketch, techniques of sketching. Charts, graphs, mathematical constructions. Presentation.

ROTH, K, *Fundamentals of Logic Design*. New York, West. Publishing Company, 1975.
Boolean algebra. Algebraic simplification. Diode gates. Karnaugh maps. Flip-flops. State assignment. Derivation and reduction of primitive flow tables. Hazards. Asynchronous segmental network design. Answers and problems.

ROTHBART, H A, *Mechanical Design and Systems Handbook*. New York, McGraw-Hill, 1964.
Mathematics and computers. Engineering mechanics. Systems analysis and synthesis. Dynamics of moving, contacting, bodies. Dynamics of materials. Machine fastener components.

ROY, R, *What is Designing? A Probe into Design Processes*. The Open University Press, Units 32–34, 1972.
Introduction: designing, how to use this probe. (Nine selected articles from various branches of design.)

ROYLANCE, T F (editor), *Engineering Design*. Conference 1964 at the University of Nottingham. Oxford, Pergamon, 1966.

Eighteen papers on general design principles and design of special machine types.

RUDD, D F and WATSON, C C, *Strategy of Process Engineering*. New York, Wiley, 1968. Creation and assessment of alternatives. Optimisation. Engineering the presence of uncertainty.

RYAN, D L, *Computer-aided Graphics and Design*. Marcel Dekker, 1979. Introduction. Principles and computerising of automated draughting. Draughting systems and programming. Pictorial representation. Computerised descriptive geometry. Graphic terminals. Vector analysis. Computer-generated charts and graphs. Sample programs and user problems.

S

SABIN, M A (editor), *Programming Techniques in Computer-aided Design*. Manchester, NCC, 1974. Language-processing. Interpreting, design. Computer graphics. Data structure. File handling. Discussion on each topic.

SACKMAN, H and CITRENBAUM, R L, *Online Planning Towards Creative Problem-Solving*. Englewood Cliffs, Prentice-Hall, 1972. Theory and method of man-computer planning. Management and project planning. Experimental analysis of interactive planning.

SANTIS, R M, *Causality Structure of Engineering Systems*. University of Michigan, 1972. Introduction. Causality on group resolution space. Connections between the concepts of causality and state. Causal properties in Hilbert resolution space. Causality, strict causality and stability.

SCHNEEWEISS, C A, *Inventory – Production Theory*. Berlin, Springer, 1977. The general linear quadratic and linear non-quadratic models. Comparison with optimal dynamic programming solutions. Comparison with deterministic approximations. Comparison with AHM-inventory models. Summary and concluding remarks.

SEELY, S, *An introduction to Engineering Systems*. New York, Pergamon, 1972. Modelling of system elements. Interconnected systems (equilibrium formulation). System response. Selected topics.

SEIREG, A, *A Survey of Optimisation of Mechanical Design*. Transactions, American Society of Mechanical Engineers, B94, 1972/2, pp 495–499.

SHIGLEY, J E, *Mechanical Engineering Design*. 2nd ed. New York, McGraw-Hill, 1972. Part 2. The design and selection of mechanical elements. Screws. Fasteners and joints. Mechanical springs. Bearings. Lubrication. Spur gears. Shafts.

Clutches, brakes and couplings. Flexible elements.

SIDALL, J N, *Analytical Decision-making in Engineering Design*. Englewood Cliffs, Prentice-Hall, 1972. Introduction to value theory. Decision theory. Introduction to optimisation. Analysis. Optimisation of linear, non-linear and segmental systems. Examples. Reliability theory. Fortran programs: 15.

SIMON, H A, *The Science of the Artificial*. Massachusetts, MIT, 1969. Understanding the natural and artificial worlds. Psychology of thinking. Science and design. Architecture of complexity.

SINGLETON, W T and others, *Measurement of Man at Work*. London, Taylor and Francis, 1971. Twenty-seven articles on management techniques and applications.

SINGLETON, W T and others, *The Human Operator in Complex Systems*. London, Taylor and Francis, 1971. Twenty-one articles on system design methods. Analytical techniques. Allocation of function. Task description. Training, job aids and maintenance. Applications.

SMITH, D A, *Automatic Generation of a Mathematical Model for Machinery Systems*. American Society of Mechanical Engineers Paper 72 Mech. 31, Meeting October 8–12, 1972. A generalised computer program. DAMN.

SMYTH, M P, *Linear Engineering Systems*. New York, Pergamon, 1972. Functions, operators and systems. Frequency representation. Laplace transform. Description and analysis of continuous lumped distributed and discrete systems. Additional topics in system theory.

SPILLERS, W R, *Basic Questions of Design Theory*. Proceedings of Symposium, Columbia University, May 30–31, 1974. Amsterdam, North-Holland, 1974. Civil engineering and architecture. Chemical engineering. Mechanical engineering. (Design aspects of parameter estimation. Fluid film bearing design. Kinematic structure of mechanisms. Electrical engineering. Socio-urban design.)

SPOTTS, M F, *Design Engineering Projects*. Englewood Cliffs, Prentice-Hall, 1968. Creativity projects, cold and hot heading, screw machine parts, machining, miscellaneous production methods, dimensions and tolerances. Strength of material, mechanical components.

SPOTTS, M F, *Design of Machine Elements*. 4th ed. Englewood Cliffs, Prentice-Hall, 1971. Fundamental principles. Working stresses.

Shafting. Springs. Screws. Bolts. Welded connections. Lubrication. Bearings. Spur gears. Bevel, worm and helical gears. Dimensioning. Engineering materials.

STARR, M K, *Decision Making*. Englewood Cliffs, Prentice-Hall, 1963.
Nature of decision. Elements of decision theory. Decision-making process (risk, certainty, uncertainty). Qualitative decision methods. Simulation methods.

SVENSON, N L, *Introduction to Engineering Design*. London, Pitman, 1976.
Society and engineering. Procedures of engineering design. Origin and identification of engineering design problems. Creativity in engineering design. Systems design. Technical and economic analysis.

T

TAYLOR, G A, *Observations on the Development of Creativity in Europe*. Thayer School of Engineering, Dartmouth College, 1967.
Need, definition, observations, 12 reports.

TJALVE, E A, *Short Course in Industrial Design*. London, Newnes-Butterworth, 1979.
Creation of a product. Methods used in form design. Form factors. The appearance of the product. Case history: chromosome apparatus.

TJALVE, E A, ANDREASEN, M M and SCHMIDT, F F, *Engineering Graphic Modelling*. London, Newnes-Butterworth, 1979.
Drawing and design. Modelled properties. The receiver. The drawing code. Drawing technique. Types of drawing.

TOMIYAMA, T, *General Design Theory and its Application to Design Process*. Tokyo, University of Tokyo, 1980.
Design theory. Design process. Representation of physical rules. Design by means of Pai-numbers.

V

VIDOSIC, J P, *Elements of Design Engineering*. New York, Ronald Press, 1969.
Design engineering and design process. Analysis and synthesis. Decision in design. Creativity. System and human factors engineering. Reliability, engineering materials and process design. Economics. Intellectual property protection. Engineering ethics. Case studies. Design projects.

VLIESTRA, J and WIELINGA, *Computer Aided Design – Proceedings of IFIP-Conference*. North-Holland, 1973.
CAD – the scope, the foundations of the manifestations. Interactive computing. CAD in system building. Methodical design as a basis of CAD. Hardware and software.

W

WELBOURNE, D B, *Problems in the Rationalisation of Pattern Mould and Die Making*. Paper for Working Party on Automation, Geneva, 14–18 May 1979.
Design and manufacture of complex 3D shapes. Designing with DUCT (Developed University of Cambridge). Computer equipment. Economics.

WOODSON, T T, *Introduction to Engineering Design*. New York, McGraw-Hill, 1966.
Problem. Planning. Need, feasibility, creativity, money, modelling, checking, optimisation, computers, communication. Students design project.

WOODSON, W E and CONOVER D W, *Human Engineering Guide for Equipment Designers*. Berkeley, University of California, 1970.
Human engineering, a design philosophy. Design of equipment and workspace. Vision. Audition. Body measurement.

WOODWARD, J F, *Quantitative Methods in Construction Management and Design*. London, Macmillan, 1975.
Quantitative methods and concepts. Introduction. Construction planning. Certainty and optimisation. Decision-making. Cost models and capital investment. Uncertainty and risk. Competitive bidding.

Y

YOSHIKAWA, H, *Multipurpose Modelling of Mechanical Systems. Morphological Model as a Mesomodel*. In Bjorke and Franksen: *System Structure in Engineering*. Oslo, Tapir, 1978.
Function model and universal model. Mesomodel. Application of mesomodel to practical problems.

Z

ZALL, P M, *Elements of Technical Report Writing*. New York, Harper, 1964.
Planning. Collecting information. Designing rough draughting (introduction, analysis, conclusion). Revising.

An Approach to Graphics System Design. Proceedings, Institute of Electrical and Electronic Engineers, vol 62, 1974.
Design of high-level general-purpose device – independent graphics system. Graphics functions. Transformations. Display file. Display processor. Outline proposal. Choice of graphic input.

Bibliography 2: articles and papers

Abbreviations

CAD CAD (GB)
CME Chartered Mechanical Engineer (GB)
Des Design, London (GB)
DS Design Studies (GB)
EMD Engineering Materials and Design (GB)
Eng Engineering (GB)
JEJ Journal of Engineering (USA)
MD Machine Design (USA)
ME Mechanical Engineering (USA)

A

ALLAN, J J, *Some research advances in computer graphics that will enhance applications to engineering design.* Computer and Graphics, 2, 1977.
Man-machine interaction. New hardware-software developments. Today's situation. Future.
ALLEN, C, *Computer-aided design, drafting and manufacture in avionics.* CAD, 6, 1974.
Automated draughting equipment developed by Ferranti. Hardware. Design. Input/output. Effect of design. Production.
ALLERTON HEGEMANN, R, *Detailers can be creative* MD, 23 July 1970.
Elusive job boundaries. Using draughtsmen's experience. Some practical suggestions.
ANDERSON ALLEN, A, *Draughtsmen can improve design.* Eng, 213, 1973.
The skilled draughtsman has a great fund of ideas and specialist knowledge of design. Is this fund being neglected or completely ignored by engineers during the course of their work?
ARCHER, L B, *Whatever became of design methodology?* DS, 1, 1979.
The three Rs. A third area of education. The vacant plot. The naming of the parts. The naming of the whole.

B

BARTEL, D L and MARKS, R W, *The optimum design of mechanical systems with competing design objectives.* JEJ, 1974.
Two journal bearing design examples. Competing objective functions in optimisation. Applications.
BENSUSAN, G and NEWBY, S, *Plan for design research.* DMG Journal, 8, 1974.
Critical article on their own theoretical approach and outputs of design research. Progress in practical design research information.

BESANT, C B, *CADMAC – a fully interactive computer-aided design system.* CAD, 4, 1972.
Reversible digitising table. Data processing console. Video display. Operating the CADMAC system. CADMAC applications. Conclusions.
BOESCH, W, *The sense of operational analysis and the use of operators for the operational diagrams of operational analysis.* Microtecnic, 4, 1950.
Six examples. (Symbolic method for studying measuring circuits, using process blocks.)
BOOKER, P. J, *Research into design activity.* EMD, 4, 1961.
Need for education in design. Design in terms of critical decisions. Time an important factor. Limited value of abstract models. Summary.
BOSMAN, D, *Ergonomic considerations in design methodology for man-machine systems.* Paper at PROMSTRA 1974.
Mental discomfort as a by-product of plant design. Partitioning of the problem. Designed and perceived architecture. Man at work.
BROOKE, E R, *Where the mechanical way still makes sense.* MD, 11 January 1979.
Adding and subtracting. Multiplying and dividing. Trigonometry Mechanisms synthesis.
BROWN, E H, *Size effects in models of structure.* Eng, 194, 1962.
Dimensional analysis. Simplifications in special cases. Failure of concrete. Yield of mild steel. Brittle fracture of steel. Fatigue. Discussion.
BRUNS, J H, *Forecasting the cost of redesign.* MD, 5 March 1970.
Generating learning curves. Using cumulative techniques.
BURY, D V, *On probabilistic design.* JEJ, November 1974.
Design inputs. Extreme value phenomena. Probabilistic design. Parameter determination. Illustration.

C

CASE, K and PORTER, M, *SAMMIE, a computer-aided ergonomics design system*, Eng, January 1980.
Operation. Applications. Tractor cab.

CAVANAUGH, W T, *How do we decide?* ME, 1967.
Trial by PR. The biased 'unbiased' expert. Academic bias. The moving target. Avoiding the three pitfalls. A new institution for technology assessment.

CHADDOCK, DH, *Sparkling ideas in the design process*. Eng, 1974.
Evaluation (PERT, PABLA). Distinct stages. Experiments.

CLARK, T S, *Ergonomics: better design or another headache?* Eng, November 1978.
Problems (ergonomic design). Development of procedures (questions, needs). Further development (refining procedures and information for trial by designers).

CLARKE, A, *Review of computer-aided design*. Eng, October 1978.
On-line in real time (OLRT) and time sharing. Interaction requirements. Graphical interaction. Presentation. Problem-solving techniques. Design program structure. File structure.

CLARKE, J A, *Choosing a design systematically*. MD, 7 January 1971.
Methods of evaluation. Figures of merit. Total evaluation.

CLAY, M, *Work study as an aid to design*. Eng, April 1974.
Basic philosophy. Rail siding (case). Enlarge horizons. Historical background. Method study. Operational research. Great coordination. Trends.

COMELLA, T M, *How to manage creativity without killing it*. MD, 6 March 1975.
What it takes to be creative. Which tasks require creativity? Are you willing pay the price? Innovation or stagnation.

CONWAY, H, *Design and produce*. Eng, 217, 1977.
Design process. Design organisation of the future. Significance of the computer. Problems of the future. Material conservation. Life-cycle costing. Flexibility in design. Industrial design.

CONWAY, H, *Getting design into focus*. Eng, 214, 1974.
Viewpoint, five proposals.

COOLEY, P, *Computer-aided drafting with refresh graphics*. Computer and Graphics, 4, 1979.
Basic draughting routines. Basic elements for 2D system. System functions. Hardware (HP 98451). Automatic dimensioning.

COOLEY, P, *Mechanical drafting on a desktop computer*. CAD, 11, 1979.

CAD on Hewlett-Packard 9845 desktop computer. Routines for mechanical draughting. Hardware. System description. Comparison to other turn-key CAD systems.

CORFIELD, K G, *Product design*. Eng, 219, 1979.
Product design management. Recommendations.

CROSSLEY, E, *A 'shorthand' route to design creativity*. MD, 10 April 1980.
Why designing is difficult. Getting a handle on complexity. Developing a catalogue of design alternations. A systems approach to design.

CROSSLEY, E, *A systematic approach to creative design*. MD, 6 March 1980.
A science of design? Analysing the design process. The method at work.

CROSSLEY, E, *Make science a partner in your design*. MD, 12 April 1980.
The relevance of science to design. Developing a catalogue of physical effects. Generating alternative design approaches.

D

DAUGBERG, R J, *Is there a better way to make engineering decisions*. Chemical Engineering, 1980.
The process. The decision-making process. Establishing objectives. Alternatives. Comparing and choosing. Evaluating risks.

DAVIS, R, *The function of creativity*. CME, 11, 1964.
The position today. Breaking through the word barrier. Motivation and creative attitude. Environment and leadership. The industrial scene. What is to be done.

DECKER, R W, *Computer-aided design and manufacturing at GM*. Datamation, May 1978.
Accelerating the design. Making tools intelligent. Making new processes possible. Pushing up productivity.

DHANDE, S G and SANDOR, G N, *Design with the law in mind*. MD, 24 August 1978.
Enlightened design practice must take into account additional criteria of adequacy for the user and the use environment. Additional criteria include identification of product, modes of operation, hazards and risks. Identification of applied standards.

DIEHL, P and HOWELL, J R, *Engineering teams: what makes them go?* ME, 96, 1974.
Models and methods. Putting the model to work. The Johari window. Applying the Johari window.

DROR, B, *Computer-aided design in Israeli aircraft industries*. Computer and Graphics, 3, 1978.
Basic concept. Historical sketch and development. Description of major CAD system in use and description of modules for preliminary design phase. Primary structure and system design phase. Detail design phase and manufacturing phase.

DUSTGATE, R, *Operators, safety-designers will carry the can*. Eng, May 1974.
Design. Discuss safety. Safe design. Rotins report. Other countries. Modifications.

E

EAGLESHAM, G, *CAD: state of the art*, CAD, 11, 1979.
Overall context of CAD/CAM. Benefits of CAD/CAM. Market trends and adoption of CAD/CAM in electronic, heavy electrical, mechanical enginering. Chemical engineering and construction industry area. Computer hardware and software.

EDWARDS, E and LEES, R P, *The influence of the process characteristics on the role of the human operator in process control*. Applied Ergonomics, April 1974.
Functions of computer and operator. Development of the operator's role. Studies of industrial operators. Implications for interface.

ELLIOT, W S, *Interactive graphical CAD in mechanical engineering design*. CAD, 10, 1978.
Interaction in 3D between designer and computer model. Modelling for analysis. Some research projects and in-use systems for 3D modelling.

EMMERSON, W C, *CAD in the motor industry – Leyland Cars*. CAD, 8, 1976.
Automobile-body design. Surface description and fairing. Computer simulation in vehicle safety. Geometric design aids. Press tool design and manufacturing.

F

FARRAR, D J, *Product competitiveness*. Eng, October 1979.
Methods of improving the relationship between company management and engineering designers.

FEILDEN, G B R, *The designer's craft*. Eng, 215, 1975.
Developing a new design. Model proof. Telling the team.

FISCHER, W E, *PHIDAS – a database management system for CAD/CAM application*. CAD, 11, 1979.

FLURSCHEIM, C, *Industrial design and engineering*. Eng, 217, 1977.
Relations illustrated by examples.

FOX, R L and GUPTA, K C, *Optimisation technology as applied to mechanisms design*. JEJ, May 1973.
Precision point syntheses. Optimal synthesis. Optimisation in kinematics.

FRENCH, M J, *Integrate it - don't add it on*. Eng, October 1975.

Undigested engineering science. Design-build-test projects.

FULLER, D, *Rating engineer performance*. MD, 24 August 1978.
Goals and priorities. What to look for: ability to adapt. Diligence, co-operation with others, quality of work, decision making, manner and appearance, job contributions, deadlines. Preparing the report.

G

GIBBS, P J, *Computer-aided integrated systems*. CME, 27, 1980.
Computer-aided design and drafting systems provide computer-stored information.

GRABOWSKI, H and EIGNER, M, *Semantic datamodel requirements and realisation with a relational datastructure*. CAD, 11, 1979.
Databank design. Semantic model level. Information modelling. Geometric model. Syntactic basic elements. Relation scheme description.

GRAYER, A R, *Alternative approaches in geometric modelling*. CAD, 12. 1980.
Alternative techniques for solid modelling. $2\frac{1}{2}$D and 3D drawings. Translation and rotational sweeping. Local modification-tweaking.

GROVER, D J, *Spatial transforms for computer-aided design*. CAD, 6, 1974.
Basic configuration of graphics system. Formation of display store. Interaction with a graphic display.

H

HARDY, P, *Real ability shows up in detail design*. Eng, 216, 1976.
Learning design principles, thinking creativity in engineering terms, group projects, assessment.

HERZOG, R E, *How engineers are coping with design imperatives 1977*. MD, 8 December 1977.
New design/redesign: common problems, universal solutions. Unique opportunities in product design.

HERZOG, R E, *New problems, new solutions in human factors engineering*. MD, March 1974.
Getting more accurate data, faster designing for more than operability.

HERZOG, R E, *Selling ideas, advertising style*. MD, 9 July 1970.
Watch your language. Three steps in selling. (Getting attention, turning attention into interest, motivating action.)

J

JACOBS, G, *Designing for improved value*. Eng, 220, 1980.

Design contribution to selling price. Cost-control action necessary. Performance and cost interactions. Direct operation cost. Trade-off. Value improvement sources. Working procedures.

JACOBSON, R A, *Resource conservation – the coming design parameter*. MD, 25 July 1974.
Problems with Serap. A time for caution. Problems in source reduction. Design for disassembly.

JAKOBSON, A, *Design with manufacturing in mind*. MD, 14 November 1974.
The communications bug. Making a virtue of formality. People moving. The program manager.

JOHNSON, R C, *Design synthesis – aids to creative thinking*. MD, 15 November 1973.
Inspiration from linkages. Mechanical devices as circuit diagrams. Synthesis by implication.

JOHNSON, R C, *Design synthesis – a new approach to engineering*. MD, 18 October 1973.
The value of optimisation. An optimum tin can. Synthesising a shutter. Designing a bearing.

JOHNSON, R C, *Design synthesis – selecting materials and dimensions*. MD, 13 December 1973.
Optimising a cantilever beam. Designing a spur gear set. Synthesising a helical spring.

JOHNSON, R C, *Design synthesis – the road to optimisation*. MD, 27 December 1973.
The systematic approach. Solving a real problem. Programmed searches. A continued approach.

JONES, J C, *Designing designing*. DS, 1, 1979.
Craft processes, design by drawing, system designing. Design thinking. Designing the design process. Design process control – review. Divergent and convergent design process.

K

KHOL, R, *The inevitable marriage of CAD and CAM*. MD, 1 June 1972.
The all-important envelope. From clay to numerical control tape. The end of the transfer-line? New way to design.

KIRKHAM, R L, *Product testing: setting measurable objectives is the key to success*. MD, 10 November 1977.
Avoiding the activity trap. Handling unexpected problems.

KOLL, *Design for production*. Machinery and Production Eng, 14 February 1968.
Detail drawings. Undesirable features in design. Conversion from functional to production design.

KROUSE, J K, *CAD/CAM bridging the gap from design to production*. MD, 12 June 1980.
CAD functions. CAM functions. Corporative efforts.

L

LAMING, L C, *Engineering science versus technology*. Eng, 216, 1976.
Time separation, integration, both together. Outlook and insight (design course, Department of Mechanical Engineering, Imperial College).

LANG, T G, *A generalised engineering design procedure*. Dissertation, Department of Aerospace Engineering, Pennsylvania State University, 1968.
Development of the design procedure. Description and discussion of design procedure. Design of submerged vehicles.

LAVOIE, F J, *How engineers use the computer*. MD, 20 August 1970.
Who uses the computer? What is it used for? What kind of facility?
Future plans. The electronic calculator.

LAXON, W R, *Selecting and evaluating CAD systems*. CAD, 9, 1977.
Objectives of CAD. Choosing the application. Hardware and software. Testing a CAD system. Costs and benefits.

LEESLEY, M E, *The uneven acceptance of CAD*. CAD, 10, 1978.
Development of CAD. Confirmed applications. Areas where CAD has made little impact. Barriers to the acceptance of CAD.

LIVESEY, R, *Computers for the design engineer II*. Eng, 8, May 1970.
The computer as design aid.

M

McCORMACK, F, *Process specifications*. MD, 8 January 1970.
Levels of control, pros and cons, the basic criteria, specific applications. A problem in procurement.

McDONALD, D J, *Human factors – the forgotten element in design*. MD, 9 September 1967.
Breaking down the task. Developing consistent methods. Recording motion patterns and times. Designing with the human in mind.

MAROVAC, N, *Interactive computer-aided 3D engineering and art design*. Computer and Graphics, 4, 1979.
DESIGNART – software. Theoretical aspects. Purpose. Database.

MATCHETT, E, *Creative design – a new approach to fundamentals*. Electronics and Power. 1966.
Commonsense approach. Mental yardstick. Mental renaissance. Maximum benefit. Total experience. Training courses. Optimum solution. TDM.

MATHUR, K, *The problem of terminology: a proposed terminology for design theories and methods*. DMG Journal, 12, 1978.
Definitions for central design concepts.

MATON, G W, *Designing for production*. Eng, October 1976.
Case: designing a new range of digital frequency meters (Marconi Instruments).

MAYALL, W H, *Let's not build barriers with words*. Des, October 1966.
Industrial/engineering design. The meanings of every term. Three main influences (elements, firms, people).

N

NADLER, G, *The effect of design strategy on productivity*. Int J Prod Res, 9, no 1, 1971.
Fundamental ideas about design strategies for increasing productivity. Comparison of results. Project illustrations. The IDEALS concept for design. Advantage of the IDEALS concept design strategy.

NAPIER, M A, *Design review as a means of achieving quality in design*. Eng, March 1979.
The nature of design. Objects of design review. Designer involvement. Design review functions. QA participation. Design review activities.

NEARY, R, *The computer as a design partner*. MD, 30 April 1970.
The optimum design. The CAD library. Getting outside help. Which CAD computer. Do you need CAD?

O

OLKEN, H, *Systems design: designing a purchased system*. MD, 3 September 1970.
Preparing a specification (label: system specification checklist and example). Application for design.

P

PARKER, R C, *Creativity, a case history*. Eng, 215, 1975.
Divergent thinking. Creativity. The individual. Problem-solving groups. Modus operandi. Control group. Noise. Gatekeepers. Results.

PEAKE, H J, *Creating a climate for motivation*. MD, 26 June 1980.
How are engineers different? Basis of motivation. Analysing the work environment. Producing a climate for motivation.

PELLMAN, R C, *Creativity by committee*. MD, 12 December 1974.
Preparing for a session. Forming the super-intellect. Priming the super-intellect. Running the session. Debriefing the session output.

PHILLIPS, M, *Design – today and tomorrow*. Eng, 211, 1971.
Government action. The industry's problems. What is design? Detail design. Design consultants. Technology transfer. Education. Tests of the trade. Status that is achievable.

PHILLIPS, R J, BEAUMONT, M A and RICHARDSON, D. *AESOP: an architectural relational database*. CAD 11, 1979.
Principles. Relations. Implementation of principles. Record structures. Dynamic storage allocation system. Graphics input, output and manipulation.

PITTS, G, *Information needs*. Eng, 214, 1974.
'State of the art' information. Problem-solving information. Technical manuals.

POND, J B, *Latest graphics systems recast CAD/CAM concepts*. Iron Age, June 1978.
Breakthroughs in interactive graphics, equipment for CAD/CAM systems. Costs and benefits. Integration of design and draughting.

R

RAIMONDI, C A, *Estimating drafting time: art, science or guesswork?* MD, 7 September 1973.
Eleven-step method, use of draughting time calculator.

RAUDSEPP, E, *Creativity is catching*. MD, 13 May 1971.
Courses in creativity. College and common sense. The crucial climate. The laboratory method. Feeding on feedback. Practice and results. Laboratory method reports.

RAUDSEPP, E, *Encouraging creative nonconformity*. MD, 12 June 1980.
Distinguishes between constructive nonconformity and recalcitrance and allow the creative nonconformist to pursue novel solutions.

RHODES, C, *Developing talent the Matchett way*. Eng, 215, 1975.
PABLA and MAUD design systems. Matter, media and meaning.

RIMMER, D J, *Design in the '80s*. Eng, 219, 1979.
Function. Cost effectiveness. Considering the environment. Reducing lead time. Some other factors.

ROBERTS, J C H, *Cost reduction starts with the designer*. Eng, October 1975.
Think twice on tolerances and finishes. Practical approach. Cost targets. Shape versus cost. Case study.

RODENACKER, W G, *Science of engineering design. Report of the VDI-meeting at Ulm*. Eng, 215, 1975.
Definition of function. Defining the design. Use of computers. Methods of solution.

S

SCOTT, A, *More value by design*. Eng, 211, 1971.
Case histories. New materials, new processes, changing economic pressures require re-examination of designs.

SHARPE, E B, *Optimising design with analog computers*. MD, 11 July 1980.
Operation. Example problem. Hybrid computers.

SINHA, S K and BUCKTHORPE, E D, *CAD technique applied to diesel engine design*. GEC J Sci Technol, 45, 1980.
Design and development of turbocharged and intercooled engines. Effect of new analytical techniques. CAD finite element technique.

Y

YRAN, K, *Designing new starts with designing*. Eng, October 1976.
Industrial design of the Philips organisation. Seven-step design track.

Z

ZELIKOFF, S B, *The obsolescing engineer*. Science and Technology, 1969.
Measuring depreciation? Upmanship. Picking a methodology. Nothing is perfect. More graphic. Educational architecture. Questionable future.

ZIMMERMAN, M D, *Packaging the human operator*. MD, 21 October 1978.
Allow for the human envelope. The flow of design. Murky models or clear examples. Left-outs versus misfits. Master models curb misfits.

ZIMMERMAN, M D *Scale-model testing*. MD, 28 May 1970.
Testing models to provide accurate data for the construction of full-size prototypes. The techniques of using models and problems associated with miniaturisation.

ZWICKY, F, *Task we face*. Journal of American Rocket Society, no 3, 1951.
Morphological thinking. Four dark clouds. Conventional methods are insufficient. Sociological and technical applications. Problems to which morphology has been applied. Totality.

Case Studies

APT power-car sole bar

British Rail's prototype Advanced Passenger Trains were designed to meet carefully formulated requirements specific to the needs of their high-speed performance. For the body structures this meant the natural vibration frequencies of the whole body shells had to be considered during the design; a factor which had not been important on earlier, slower trains. In addition the APT had to meet the new international requirements for structural resistance to longitudinal buffing loads.

On the more-lightly-loaded trailer cars, the passenger-carrying vehicles, the natural-frequency requirements were fairly easily met and only the buffing cases caused problems for the designers. The extra mass of the equipment in the power cars depresses the natural frequency to such an extent that the structural design of the basic shell became mainly a problem of meeting the stiffness requirements within the severe weight restrictions.

This case study looks at the evolution of the design of one of the main structural members of the power car between 1973 and 1979.

1 *Prototype Advanced Passenger Train*

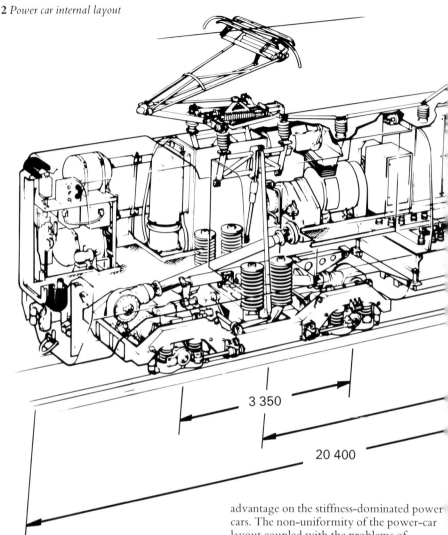

3 350

20 400

Overall configuration

Aluminium has a higher strength-to-weight
ratio than steel and was the obvious choice
for the strength-dominated structure of the
trailer cars. In addition the relatively
uniform cross-section of the trailer cars
meant that good use could be made of
aluminium extrusions to reduce assembly
costs enough to offset the higher basic
material price. However, the
stiffness-to-weight coefficients of
aluminium and steel (and of most common
structural materials) are almost identical,
which removes aluminium's weight

advantage on the stiffness-dominated power
cars. The non-uniformity of the power-car
layout coupled with the problems of
welding complex aluminium assemblies
gave steel a large cost advantage. The other
important factor to be considered was the
manufacturer. British Rail Engineering Ltd
had a long history of building steel vehicles
but none of welded aluminium ones.
Therefore the material choice was steel and
with only six prototype cars to build the
choice was limited still further to standard,
easily available stock steel.

The early power-car equipment layouts
required the vehicle to be 21m long with a
bogie spacing of 14m and a bogie wheelbase
of 3.5m. (These dimensions later became
20.4, 13, and 3.35m respectively, figure **2**).

144

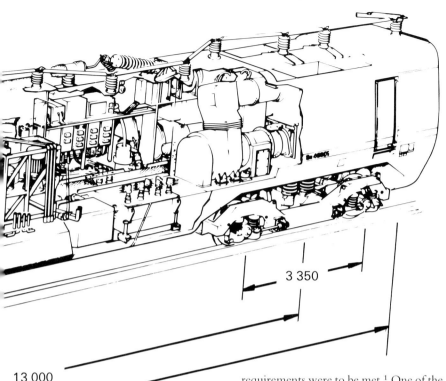

3 350

13 000

The four $1\frac{1}{4}$-tonne traction motors were mounted on the body above the floor to cut down the bogie mass. To keep the vehicle's centre of mass as low as possible, the heavier items of power control equipment – the transformer, main chokes, and their cooling systems; some 12 tonne in all – were mounted below the floor in the middle section of the vehicle. The rest of the expected equipment and structural masses were distributed along the vehicle length and a rough structural configuration sketched in. This was translated into a numerical model for a basic finite element analysis by computer. By applying the worst buffing load cases of 2000kN compression and 1500kN tension at the coupler level, and deriving natural-frequency modes from the analyses, the suitability of various structural designs could be assessed. It soon became apparent that it would be necessary to make a very efficient use of the mass available for basic structure if the stiffness and strength

requirements were to be met.[1] One of the most important members was revealed as the sole bar. This is the member which runs along the junction between the wall and the floor. It carries a major proportion of the buffing loads and makes a significant contribution to the lateral stiffness of the vehicle. This is the most important mode to match the dynamic requirements of the bogie attachments. The target lateral frequency was 12Hz with a $10\frac{1}{2}$Hz requirement vertically. After some detailed investigations based on the tentative information then available an overall structural configuration to meet these requirements was derived. In this the sole bar needed to have a cross-sectional area of 4000mm^2, provide bending continuity between framing members in the wall, floor, and skirt, and be stable under compressive loads between framing members. In addition it had to incorporate some form of joint to allow the assembly of the walls to the floor during construction.

[1]The basic structure design mass was limited to 10.5 tonne with another 2 tonne available for local brackets and mounts.

Space restrictions

The exterior limitations of loading-gauge clearance define the outer profile of the vehicle. The layout of the equipment is severely constrained by the sheer size of some of the items and the requirement to provide a corridor for crew access and emergency passenger use through the car. The remaining spaces between the equipment and the outer profile were available for the structure. At the sole bar this space was limited by the secondary springs on the bogie, which protrude through the floor, the bogie clearance envelope, the traction motor mounts and the outer profile (see figure 3). Similar restrictions limited the floor depth to 150mm (upper limit 1400mm above rail level) and wall structure (skin plus framing) to 70mm.

Vehicle assembly sequence

At an early stage the various complexities of floor connections, space restrictions, and structural loads led to a decision to split the floor into five main subassemblies. These were a central section, mainly of thin sheet construction, including the underfloor bulkheads and lower sidewalls; two over-bogie sections incorporating the complex, heavily-loaded bogie connections and the various holes, wells, and mounts for the mechanical transmissions; and two end structures carrying the very heavy structure necessary to distribute the buffing and coupler loads into the body shell. These were to be put together into a single floor assembly to which the walls would then be added. To avoid the problems of making large butt welds on the floor assembly, when access would be restricted, it was further decided to concentrate on sole-bar designs with sections already made up to the full length of the car. These sections could be added to the floor assembly, be incorporated into the wall assemblies, or have part on each to form the joints between the two.

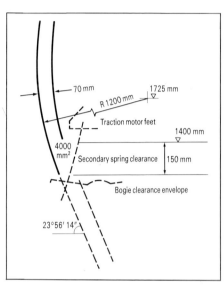

3 *Sole bar space limitations*

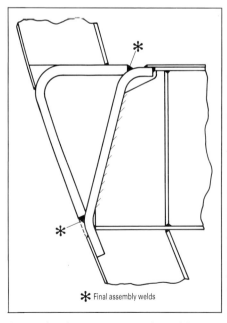

✳ Final assembly welds

5 *Bent plate design, one-piece subassembly*

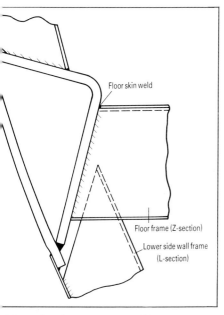

4 Bent plate design, main assembly joint

Labels in figure 4:
Floor skin weld
Floor frame (Z-section)
Lower side wall frame (L-section)

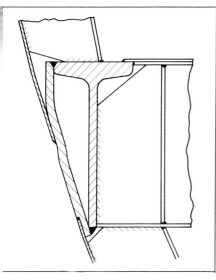

Rolled steel design, T and plate

Possible solutions

From the shape of the space available and the requirement to provide bending continuity of the frames, it was apparent that the sole-bar section must be a single solid section or a triangulated hollow section. The components of the section could be formed from flat plate or bar, from standard rolled sections, or from special extrusions. (This latter route was explored since aluminium-alloy extrusions had proved so effective on the trailer cars. The costs were known to be high and uneconomic for the prototype build, but an effective solution could have carried over into the production run and may have proved cheaper in the long term.)

A number of plate solutions were explored using plates formed in a press then butt-welded to form 20m lengths. Figure **4** shows a typical two-part solution, with the main longitudinal welds being made at the final assembly stage. The problems of accurately positioning, clamping, and welding these final joints did not favour this solution. In addition the large dirt trap behind the wall skin would have caused a corrosion problem. The design shown in figure **5** is a typical one-piece solution which probably would have been included in the sidewall assembly. Its main drawback is a requirement for extreme accuracy of manufacture both of the sole bar itself and of the floor in order to make an adequate weld on the edge of the floor skin. In addition a high positional accuracy of the floor and lower sidewall framing was necessary to make the frame joints. (The version shown is for a shallower floor at 1350mm above rail level.)

Most of the rolled-steel sections available are too large to be considered here and most of those remaining were found to be of the wrong proportions to fit the space available, although many configurations were explored. One interesting possibility is shown in figure **6**. This combination of a plate and T-section would have formed part of the floor assembly. The lap joint for the floor skin and the vertical frame weld reduced the need for accurate floor assembly and gave a convenient vertical datum plane for frame positioning. The final assembly with the wall was less satisfactory although all the welds would have been small. Again there is a dirt trap problem. The design shown in figure 7 proved to be much more satisfactory. This was to be made by welding a machined plate across the legs of a channel section which had been machined away to give the required shape. The main section was to be assembled with the sidewall where the horizontal top surface provided a jigging position for the frame members. The small floor-edge angle could be lap jointed with the floor skin to align it with the jigged positions of the floor frame ends. Thus on final assembly the joint was self-aligning horizontally and the sidewall would have considerable vertical stiffness but some horizontal flexibility, while the floor, at this stage, would have lateral stiffness but be flexible vertically allowing some adjustment. This design had considerable attractions and became the baseline design to which the others were compared. It was appreciated at the time that machining away parts of a rolled section would probably introduce some curvature in the section, which would be compounded by the non-symmetrical positions of the two large assembly welds. It was decided to allow for a straightening procedure to be applied to each part of the sole bar before welding up the full 20m length.

The possibilities of extruded-steel sections were explored after the baseline design had been evolved and reflected its influence. The overriding restriction on extruded-steel sections is the limitations on size. Because of the very high pressures necessary to cause the steel to flow a small die is used to stop the total loads becoming excessively high.

At the time that this work was done, late 1974, the die size was limited to 120mm

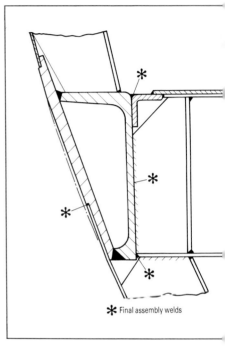

✻ Final assembly welds

7 *Rolled steel design, channel and plate*

diameter. Figure 8 is a direct copy of the baseline design using extrusions. The limitations of die size result in the obvious problems of multiple welds, difficult clamping problems, and an altogether messy assembly. Figure 9 is a better solution with a similar overall shape, incorporated into the floor assembly this time and, by using a plate insert, with considerably cheaper extrusion costs. This was probably the best of the extrusion suggestions, but costing produced such a high figure that it seemed unlikely to be economic even over a reasonable production run (60 trains at 20 per year.)

The baseline design was chosen and subsequent design work over the next two years was based on it. As the rest of the structure evolved it became necessary to make a number of local modifications but

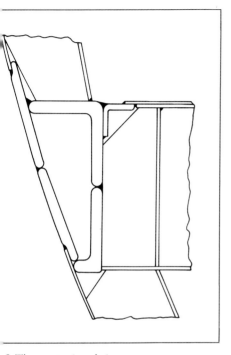

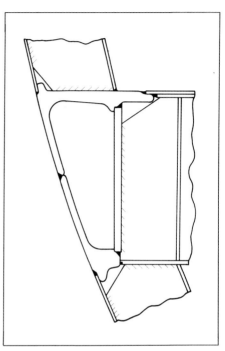

8 *Three-extrusions design*

9 *Two-extrusions-and-plate design*

the basic design remained valid. Above the buffing structure and at the ends of the lower sidewall it became apparent that large longitudinal structural connections with the sole bar would be necessary. This was accomplished by replacing the outer plate locally with a thicker plate shaped to match the adjoining members. Use was made of the tapered lower part of the sole bar when bogie slinging points, for lifting purposes, were required. A tube, inserted across the section and welded in place, provides an attachment point for a shackle with sufficient vertical strength and without much change in the properties of the section.

The section was produced and incorporated into the prototype power cars now in limited service with British Rail. The design proved to be satisfactory for assembly purposes and the vehicle passed the

structural testing programme meeting all its strength and stiffness requirements in early 1978. The actual mass of the complete structure was 13 tonne with the two sole bars accounting for 10% of this figure.

As anticipated, some problems were encountered with straightening the sections after machining and welding and the costs of this procedure would have been prohibitive in full production. Thus when the production power-car design was started late in 1977 this problem led to a re-examination of the design.

Production design

There are a number of fundamental differences in layout and configuration between the prototype and the proposed production power cars, but the sole-bar requirements remain relatively unchanged, except for local attachments.

One design using a standard bulb-flat section had been discarded at the prototype stage since the standards were being revised and it was thought that this type of section, not extensively used on the railways, might well become obsolete. However, the new standard contains bulb-flats and the design was resurrected for examination. In combination with a standard angle, slightly machined, it proved possible to produce the same basic shape as the previous sole bar but with the welds more symmetrically disposed about the section centroid. Consultations with the works production planners produced a suggestion to eliminate the small floor-edge angle with a little care in manoeuvring the sidewalls into place. The economics of production quantities also allowed a change in the sidewall frames from a stretch-formed angle to a pressed 'top-hat' section with less depth. This gave room to allow the floor skin to overlap the sole bar and thus take up the works suggestion. The resulting design (figure **10**) contains all the useful features of the prototype with a simpler, hence cheaper, floor design. Discussions with the section producers have shown that, with only a small additional cost, an extra roll stand at the manufacturing stage can be used to slightly curve the bulb flat to match the outer profile.

In 1979 some 5m lengths of this design, using a standard bulb flat, were welded up and had no measurable curvature, thus confirming the design assumptions. Due to changes in the overall train configuration there will be two types of steel vehicles: power cars and auxiliary equipment/van vehicles. Both will use variations of the new sole bar design when they go into production in the mid-1980s.

NOTE Since this unit was written the APT prototypes have completed a long evaluation programme. Although the production version in the form described

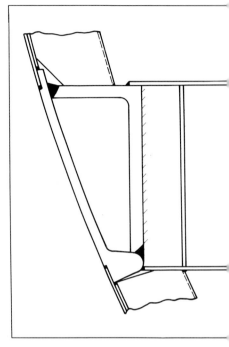

10 *Production version*

here has been cancelled, most of the lessons learned from the project will be applied to new trains due in service around 1990. In particular the structure evolved for the APT power cars will be adapted for lightweight passenger locomotives now being developed.

References
Boocock, D and King, B L, 'The development of the prototype Advanced Passenger Train', in *Proceedings of the Institution of Mechanical Engineers* vol 196, no 6, 1982.
Boocock, D and Newman, M, 'The Advanced Passenger Train', in *Proceedings of the Institution of Mechanical Engineers*, vol 190, no 2, 1976.
Ledsome, C, 'APT: British Rail's train for the future', in *Engineering*, February 1981.

Doubles detection unit

When counting, sorting, and dispensing small pieces of paper, such as playing cards, lottery tickets, computerised payment forms, there is a need for a reliable means of insuring that only single sheets are counted, ie a detector capable of distinguishing between no paper, single sheets, and multiple thicknesses. Furthermore it is vital that such a detector, whether it is used as part of a thickness controller, or as part of a counting and sorting system, is accurate. The accuracy required of such machines can be of the order of one miscount or less in 10,000,000 events.

The case study below is described in the context of a low-cost, medium-volume (1000 units/year) sort/count machine that is to be used with low-security printed forms. Initially the machine was aimed at the United States. However, in the longer term the machine is projected at the worldwide market with associated variations in form size, shape, and thickness.

Specification

The initial specification included the following points:
1 A unit to detect:
  a no paper
  b single sheets
  c multiple sheets
2 The single-sheet condition must be capable of resolving:
  a corner folds above 2mm in length
  b tape 0.075mm thick and 5mm long stuck to the surface of the document
  c Z-bends, half bends, and other gross distortions of the paper
3 The dimensions of the documents are:
  a 50–100mm
  b 100–200mm
  c 0.08–0.12mm
The forms are transported in the direction indicated (short-edge feed) at a speed of 2m/s. The maximum feed rate is eight documents/second.
The temperature range is nominally

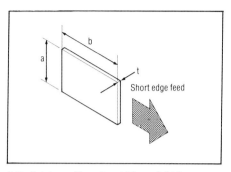

1 *Definitions of length, width, and thickness*

0–55°C although any electronics will be located in an area which might rise to 70°C. The projected cost of the complete unit is £50 at quantities of 1000 per year and £10,000 is available for tooling.

The life of the machine is to be five years, maintained by service personnel every three months, or 6,000,000 forms, whichever is sooner. It is possible to ask the operator to perform simple cleaning functions at the

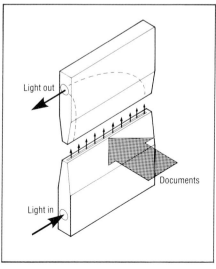

2 *Holes and tears detector using fibre-optics*

start or end of each day; however, this is undesirable.

A machine jam is different from a miscount, but more than one jam a week is unacceptable. The system must also be tolerant of large thickness changes such as paper-clips, gulp feeds, and sellotape.

The nominal design rating is 50%: the machine is expected to be used for four hours per eight-hour day. However, it is predicted that operators will not take notice of this. Therefore any device must be capable of ten hours continuous operation five days a week.

Implementation

There are a number of accepted ways of measuring document thickness. Two methods dominate the market:

1 Light (visible or infra-red) is shone on the paper and the transmitted or reflected energy measured.
2 The document is passed through a pair of rollers arranged like a mangle so that one roller deflects and this deflection is measured (see figure 3).

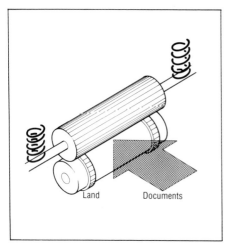

3 *Mangle type of thickness detector. The lands are about 25% of the thickness of a typical document*

Initial solution

The prototype design included a pair of mangle-type rollers and also a fibre-optic fishtail to measure transmitted energy. The latter is primarily for holes and tears detection and is neglected for the purpose of doubles sensing.

The original doubles detect unit made use of fibre-optics, but not in the standard manner. It was arranged thus:

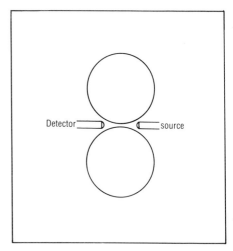

4 *Initial sensor design. Transmitted light increases with thickness*

The two lands on the lower, reference, roller allowed a clearance of 0.02mm between the two rollers in the portion where the optical sensors were placed. This was to provide a reference level for the electronic signals being processed on the main printed circuit board (this was then tolerant of long-term intensity variance and some dust build-up).

As the paper passes through the rollers the upper roller is forced to deflect, so causing an increase in the gap width and hence an increase in the light falling on the receiver. Thus for a single document the signal would

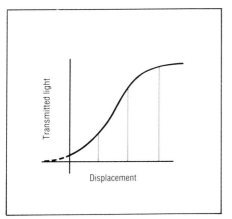

5 *The increase in transmitted light is not directly proportional to the displacement of the sensor roller*

increase by a factor of three or four relative to the reference; for a double document the increase would be perhaps nine or ten times the base level.

This non-linearity is due to the position of the emitter relative to the sensor roller. As the sensor roller deflects upwards more of the optical fibre is revealed so that the increase in received light is more sinusoidal than linear.

Because there was a suitable light source available and some investment in fibre-optic

technology had already been made in the form of the holes and tears sensor, this was an admirable, low-cost solution to the problem as then perceived.

Limitation of initial design

When extensive tests were carried out at full speed it was discovered that the shock of accelerating paper from standstill to 2m/s caused, among other things, ink and dust to be deposited on to the sensors. These rapidly became too dirty to provide reliable resolution between documents. The operator could not easily gain access to the sensor pairs and the dust was electrostatically charged so could not be blown away. The proposed solution was to fit small pipes on the end of the optical fibres so that the back pressure inside the pipe would not allow dust to enter. This will work provided the internal diameter is less than about one third of the length of the pipe. No gradual deterioration of the signal was noticed during development work with sample forms prior to an on-site demonstration with a company in New York. Unfortunately during the first week of trial the machine suffered from frequent stoppages. It was found that large chunks of paper and pieces of fibre were being shaken loose by the action of the acceleration rollers. These pieces were typically 1–5mm in size and some of them managed to block the light

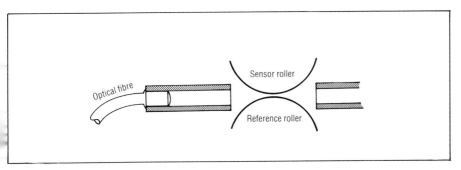

6 *Modifications made to the optical fibres to reduce the build-up of dust particles.*

passing between the sensor and receiver, causing a failure of the doubles detect system. Whenever this happened the machine stopped to avoid the possibility of passing double documents through the system.

The machine was returned to the United Kingdom and work commenced on alternative designs. During the post-mortem it was suggested that the document samples held by the company had been sorted so many times that almost all of the dust and loose particles that were originally attached to the documents had been shaken off long ago, so they no longer provided a good approximation to real life. The test documents held on site do not always represent the forms found in the typical office. It is not unknown, for example, for a machine to sort 10,000 mixed documents in a few minutes and then half a dozen engineers spend ten times as long deliberately mixing them up again!

Alternative solutions

The fibre-optic method was held in abeyance for a time and the ideas that had been generated by individual members of the design team were extended to a formal brainstorming session. The mangle rollers were defined as an essential part of the system so some method of detecting the vertical movement of the upper roller was sought. The criteria were:

1 There should be no mechanical contact between sensor and roller.
2 There should be at least two channels.
3 The response time should be less than 1μs.
4 Each channel should retain as much of the existing electronics as possible.
5 The cost to be less than £10 per channel for components.

All of the other methods suggested – and these included microwaves, radio waves, ultrasonics, beta radiation, lasers, piezo crystals, and optical devices – the most promising seemed to be the use of an

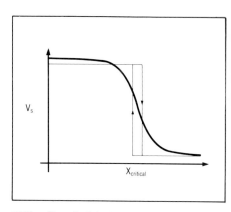

7 *The effect of a Schmitt trigger is to reduce bounce at the change-over point*

inductive proximity switch mounted above the sensor roller.

Most inductive switches are logic compatible and have an integral Schmitt trigger. This trigger causes the change-over to be rapid and precise and reduces bounce.

Some types of proximity switch, known as three-wire types, do not include such triggers and have a finite slope on the distance current response graph. Thus there is a degree of sensitivity inherent in the sensor. The current output of the sensor is dependent upon the distance of a metal target. This dependence is precisely what is required of the doubles detector. The sensitivity of the device – the slope of the curve – may be varied by altering the value of the load resistor (R_2). Proximity switches are current devices; however, the signal analysis must be carried out in terms of voltage. For this reason a relatively high-load resistor was chosen; greater than 10kΩ and this was varied until the voltage change across the load was 2.5V/mm. This voltage was fed to a voltage follower and then through an analogue circuit that provided an auto reference against the base level (note absent condition) so that if the supply voltage drifted, or something else affected the zero position of the unit over a long time-scale (about a minute), the signal would not be degraded.

A typical oscilloscope trace for a damaged document that had been repaired might look like the one in figure **8**.

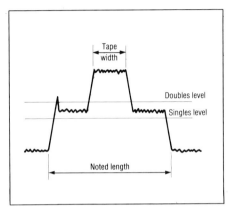

8 *An oscilloscope trace of a typical damaged document. The spike at the left-hand edge is due to a corner fold and the lump in the middle is due to tape stuck on to the surface*

The analogue signal is converted to digital using two comparators (a comparator is an electronic device that compares two input voltages). If input A is higher than input B, the output is high (5V in the case of TTL); if B is higher than A, the output is low. If the output is high, the comparator is said to be 'set'. The comparators in this device compare the raw analogue against fixed voltages called the 'singles level' and the 'doubles level'.

When the analogue voltage is below the fixed singles level, the processor 'knows' that no document is present. However, when the signal voltage increases beyond the singles level, the comparator output changes state and the processor knows that a document is present. Similarly with the doubles level: an analogue voltage above the doubles level indicates to the processor that the paper passing through the detector is too thick. Depending on how long the doubles level is exceeded, the processor will decide that a document has tape stuck on it or it was indeed two sheets of paper passing through the detector simultaneously.

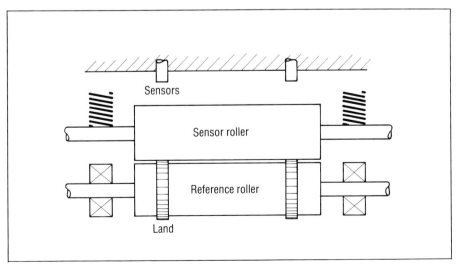

9 *Arrangement of the detector system*

It is possible to select the size of flow that
will be ignored by the processor so that a
varying proportion of documents will be
regarded as unusable depending on the
requirements of the user.

A prototype detector was built and tested.
The results were very good and proved that
the system would work.

The price, at £6 per sensor, was satisfactory
and a second machine was sent to New York
for testing. Again it worked satisfactorily.

A further dozen detectors were made up
and pre-production testing commenced. It
was noticed that the sensitivity of some of
the units seemed to vary with the time of
day and further investigation showed that
this drift was due to temperature effects.
Further tests were carried out and it was
discovered that some sensors were very
susceptible to temperature changes while
others were not. Another brainstorming
session was held to discover a means to
compensate for this variation in sensitivity
and the manufacturers were contacted to try
to produce a 'special' that would have
guaranteed temperature stability.

While this was being investigated, a third
solution was engineered. In the past the
displacement of the roller was measured by
a sort of cam follower arrangement on top
of the sensor roller.

The armature of a linear voltage
displacement transducer (LVDT) was
attached to a sprung roller follower. As the
sensor roller moved up and down the
armature followed the movement.
This system was implemented in the
pre-production models but is fundamentally
unsatisfactory:

1 The electronics are relatively complex:
requiring a high-frequency oscillator,
demodulator, and amplifier in addition to
the electrical DC circuitry.
2 The cam followers tend to pick up dirt
and grease from the sensor rollers.
3 It is mechanically complex and increases
the audible noise levels at higher
frequencies.

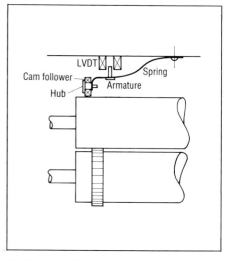

10 *The system became much more complex
when the LVDT units were used*

4 It is expensive.
5 It is complex to service.

Eventually an inductive system was
designed that was insensitive to temperature,
had a wide variation in sensing range, and
was also cheap to produce.

The electronics for this were fairly
complex, but the mechanical hardware
consisted merely of a coil and former
suspended above the sensor roller.

When the coil is energised with a short
(1μs) voltage pulse, the voltage takes a finite
time to decay because of the inductance of
the coil. If there is metal in the
neighbourhood, then eddy currents are
produced in this metal. These absorb energy,
causing the pulse to decay even more
rapidly. The new system made use of this
phenomenon by defining two sampling
times at known intervals after the end of the

156

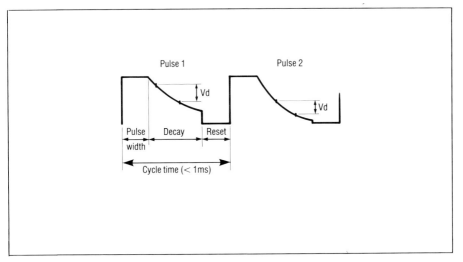

11 *An oscilloscope trace of the voltage decay system. The first pulse represents an occasion when the metal target is distant; the voltage difference Vd is large. The second pulse occurred when the target was nearer; the decay is much more rapid due to the eddy currents and so the measured voltage Vd is different. The size of Vd is a direct measure of the distance of the target*

energising pulse. The difference between the two voltages at these times gives a measure of the rate of decay of energy and hence the proximity of ferro-magnetic material.

The time constant of the coil is made to be of the order of $300\mu s$ so that decay is essentially complete when the next energising pulse is due after 1ms or less.

The pulse lasts a few tens of microseconds and the first sample is taken just after the energy pulse has ended. The second sample is taken about $100\mu s$ later. The precise times and time constants depend on the hardware available and the sensitivity required of the detection unit. However, a very sensitive system can easily be constructed with a fast response time. This too has suffered from teething problems and was rejected after showing some initial promise.

The decision was taken eventually to use the tried and tested LVDT system and review other designs as they were proposed.

This is a fairly representative example of how the detailed design of a machine element will go through successive iterations before sufficient information is gathered about the various solutions. In fact time pressure became very great during the development of the detector and this was one of the major factors in the final decision to adopt a system that did not fulfil the original specification, that is to say, that although the system was known to be less than 100% reliable and it had a number of undesirable features, greater emphasis was placed upon the degree of technical risk involved. This may be summed up by saying that, in general, if a design appears a bit suspect it will often be rejected in favour of a more tried and tested solution.

Case Study by J Vogwell
University of Bath

CAD on the floor

The manufacture of industrial flooring is extremely competitive, especially in times of economic recession when companies are fighting for fewer orders. A British company, which is recognised as a market leader and is determined to maintain its position, asks whether its product is the most efficient available and whether the technical data are displayed to their best possible advantage. This study describes an optimisation solution which has enabled the company to rationalise its product range and improve its competitiveness. In addition the material content has proved to be well suited to undergraduate computer-aided design analysis projects.

Overall configuration

Industrial flooring must be designed to be safe, rigid, and reliable under different loading conditions. It has a variety of constructional applications ranging from pedestrian walkways to supporting heavy loads such as moving vehicles or bulk storage. It is to be found extensively in factories, power stations, ships, and oil-rig platforms and is produced for a market which has a multi-million-pound annual turnover.

Companies produce flooring in a wide variety of geometric forms, using different manufacturing processes and materials such as steel, aluminium, glass-reinforced plastics, and other composites. The most widely used form, however, is open-tread steel grating, which comprises lengths of rectangular-section bar secured at an equal spacing pitch by forge-welded transverse rods. This form of flooring is economical to manufacture using special purpose machines and the material cost is the

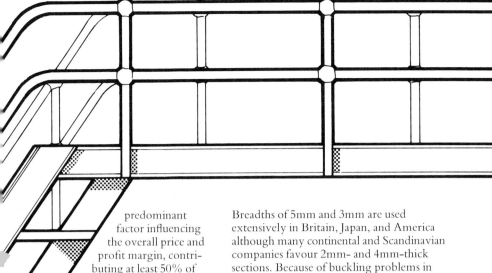

predominant factor influencing the overall price and profit margin, contributing at least 50% of the final cost.

In the design of flooring the size of section to be specified is dependent upon the load to be supported and the span which is to be bridged in satisfying a customer's safety and rigidity requirements. Usually these requirements comply with industrial flooring standards and codes of practice, but since these tend to vary significantly from country to country the result is that manufacturers in many parts of the world design to different recommended standards.

Obviously a manufacturer who wants to be internationally competitive must base his quotations on designs which just fulfil the minimum customer requirements. To design to a more restrictive standard than is applicable, or to incorporate further safety margins upon an already conservative design recommendation, will inevitably lead to a loss of orders or reduced profit margin. It is essential, therefore, that a company's product range is efficient and that the technical data enables an optimum section to be identified as simply as possible for a given specification.

Description

The terminology used in the open-tread metal-flooring industry is given in figure 1, which also illustrates how the panels rest upon the supporting structure. Standard sections are used which have depths ranging from 20 to 60mm in 5mm increments.

Breadths of 5mm and 3mm are used extensively in Britain, Japan, and America although many continental and Scandinavian companies favour 2mm- and 4mm-thick sections. Because of buckling problems in manufacture the depth/breadth ratios are limited; for the company considered in this article this was 15:1. In addition panels are produced in a range of spacing pitches and in applications where no gaps at all are permitted it is usual to tack-weld a thin sheet to the top surface. Where good grip is required the top surface of the sections is provided with serrations.

Flooring terminology

Panels are manufactured in sheets 1m wide and in lengths of 6.5m. In the erection of industrial flooring, panels are cut to the required length of span and secured to the supporting structure using brackets. Panel widths are added as required and clipped together. Steel is the metal most commonly used and is galvanised in applications where there will be exposure to a corrosive environment. Aluminium is sometimes used when lightness is required, but aluminium panels are fabricated from extrusions rather than welded.

Design considerations

About 80% of flooring is required for pedestrian loading and the designs are restricted by both strength and deflection constraints. Flooring which is to support bulk loads – which may be stationary (in storage applications, for example) or dynamic (such as vehicle wheel loads) – is limited by strength considerations alone. The British Standard 4592:1970 is the most

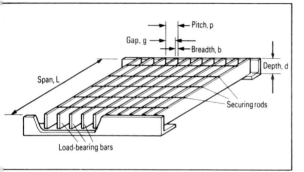

1 *Flooring terminology*

comprehensive and widely used standard available and sets out minimum standards for steel used and critical design loads as well as performance standards.

BS4592 states that the design for pedestrian loading should be based upon a uniformly distributed load of 5kN/m² and this must not produce a deflection greater than 10mm or 0.5% of span (whichever is the lesser). Also the maximum allowable spacing gap is 38mm and the minimum quality of mild steel which may be specified is grade 43A to BS4360, which has a permissible design stress of 165N/mm². It is believed that the 5kN/m² loading corresponds to the pressure which the weight of an average man would apply to a square foot area. For bulk loading it is necessary to consider the total load and corresponding contact areas and ensure that the permissible stress has not been exceeded.

Because most panel sections are just resting on the supporting structure, it is usual to design upon the most adverse assumption that panels are simply-supported beams and that transverse rods do not provide any significant resistance to bending.

Statement of the problem Preliminary discussions with the flooring company's personnel indicated a number of problem areas. Firstly, the technical brochure

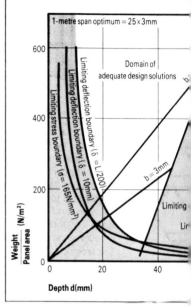

2 *Effect of span on optimum section*

presentation of flooring data was considered by both technical and sales engineers to be complex to use, and customers unfamiliar with the product had even greater difficulty. Secondly, there was concern about the suitability of the product range because some panel sections were rarely required. Thirdly, it was felt that export orders were lost because some customers, although insisting upon safety, were prepared to accept greater flexure than that permitted by BS4592. Fourthly, there was confusion over which grade of material ought to be specified. Some competitors were using better-quality alloy steels and it was important to clarify whether this was advantageous. Finally, it was necessary to consider whether the product might be improved by changing to an alternative form of construction, such as a pressed egg-box shape, or to a different section shape such as an I section.

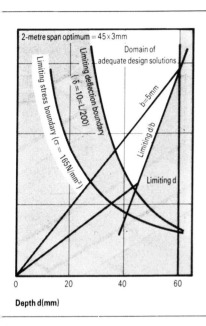

2-metre span optimum = 45×3mm

Domain of adequate design solutions

Limiting deflection boundary (δ =10=L/200)

Limiting stress boundary (σ = 165N/mm²)

b=5mm

Limiting d/b

Limiting d

Depth d(mm)

0   20   40   60

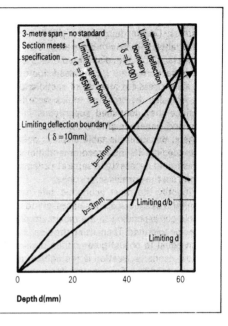

3-metre span – no standard Section meets specification

Limiting stress boundary (σ =165N/mm²)

Limiting deflection boundary (δ =L/200)

Limiting deflection boundary (δ =10mm)

b=5mm

b=3mm

Limiting d/b

Limiting d

Depth d(mm)

0   20   40   60

Summary of objectives

The primary aim of the exercise is to improve the competitiveness of the open-tread flooring company. This necessitates improving the technical presentation so that the geometry of the economic panels can be more readily extracted under both pedestrian and bulk concentrated loading. Also alternative section shapes are to be considered to establish if any gains are to be achieved and whether the overall variety of the products in the range can be reduced.

Pedestrian loading theory

The cost structure for fabricated flooring contracts comprises roughly 50% material cost with labour costs below 10%. Therefore an optimum design with regard to cost implies using a minimum weight of material.

Considering an arbitrary portion of flooring we find that:

$$\text{material weight/panel area} = \rho g \,\frac{bd}{p}$$

equation 1

where ρg is the weight density, d the section depth, b the breadth, and p the pitch. To obtain an optimum, all parameters in the numerator must be a minimum and the pitch must be a maximum. Equation 1, according to Johnson,[1] is the primary design equation. In addition we have three subsidiary design equations from two deflection constraints; and a stress constraint may be derived as follows:

$$\text{maximum centre deflection} =$$
$$\frac{5\,\omega p\,L^4}{32\,E\,bd^3} \quad \text{or} \quad \frac{L}{200} \quad \text{(for L} \leqslant 2000\text{mm)}$$

equation 2

maximum centre deflection =

$$\frac{5\,\omega p\,L^4}{32\,E\,bd^3} \text{ or } 10mm \text{ (for } L \geqslant 2000mm)$$

equation 3

maximum stress =

$$\frac{3\,\omega p\,L^2}{4\,bd^2} = 165\,N/mm^2$$

where ω is the load per unit panel area.

equation 4

Finally, because of manufacturing constraints and standard sizes, there are what Johnson calls the Limit Equations: $20 \leqslant d \leqslant 60$; $b = 3$ or 5; gap $\leqslant 38$; and $d/b \leqslant 15$. Rearranging equations 2, 3, and 4 in terms of the bd/p quantity then substituting into equation 1 enables the graphs of figure **2** to be constructed. Such a presentation helps to explain the situation and identify the salient design constraints. Optimum sections can be identified for each span given and clearly it

is the deflection constraints which are critical thus indicating the futility of using steel with a higher permissible stress.

Tabular presentations

Safe stress. Design data was originally presented in tabular form with separate tables for each pitch (table 1). Uniformly distributed loads and centre deflections were given based upon the permissible stress for a range of sections and spans. Actually the brochure tables were less clear since they contained superfluous information and panel weights were to be found elsewhere. Because this table displayed deflections exceeding BS4592 recommendations, a line was found necessary to separate adequate designs from the unsatisfactory ones.

To identify the most economic safe panel section for a desired span it was necessary to

Table 1 Safe loads

Bar section	Weight N/m²		Clear span (mm) 1250	1500	1750	2000	2250	2500
U = Uniformly distributed load (kN/m²) D = Centre deflection (mm)								
50 × 5	577	U	63.3	44.0	32.3	24.7	19.5	15.8
		D	4.2	6.1	8.3	10.9	13.8	17.0
55 × 5	528	U	53.2	36.9	27.1	20.7	16.4	13.3
		D	4.6	6.7	9.1	11.9	15.0	18.6
50 × 5	480	U	44.0	30.5	22.4	17.1	13.5	11.0
		D	5.1	7.3	10.0	13.1	16.5	20.4
45 × 5	432	U	35.6	24.7	18.1	13.9	11.0	8.9
		D	5.6	8.1	11.1	14.5	18.4	22.7
45 × 3	259	U	21.3	14.8	10.9	8.3	6.6	5.3
		D	5.6	8.1	11.1	14.5	18.4	22.7
40 × 5	384	U	28.1	19.5	14.3	11.0	8.6	7.0
		D	6.3	9.2	12.5	16.3	20.7	25.5
40 × 3	230	U	16.8	11.7	8.6	6.5	5.2	4.2
		D	6.3	9.2	12.5	16.3	20.7	25.5
35 × 5	336	U	21.5	14.9	11.0	8.4	6.6	5.3
		D	7.3	10.5	14.3	18.7	23.6	29.2
35 × 3	201	U	12.9	8.9	6.6	5.0	3.9	3.2
		D	7.3	10.5	14.3	18.7	23.6	29.2

use the table corresponding to the maximum pitch that could be tolerated. Then, using the clear span column equal to or just greater than required, the most economic section was the lightest one above the solid line. For example, for a span of 1850mm, look down column 2000 and select 45 × 3mm section.

Safe stress and safe deflections. The table of safe stresses was produced on the basis that the stress was the most important design constraint and deflection was considered as an afterthought. This represents a considerable improvement because UDLs (uniformly distributed loads) and deflections are based upon stress and deflection limits – whichever is critical. Equations 2, 3, and 4 have been used with the result that the stress-dependent spans remain unaltered and the more relevant deflection-dependent spans have changed. However, the methods of using the tables are similar and for the 1850mm clear-span example it is again necessary to look down the 2000mm column (table 3). Adequate sections to BS4592 are those exhibiting a UDL of at least 5kN/m² and consequently we again choose the 45 × 3mm section as the most economic. The major advantage with this form of presentation is that the table is now general and may be used directly for any design UDL.

Safe spans. The safe stress and safe deflection table still has some disadvantages since optimum sections may be overlooked due to rounding up to the next largest clear span. Also separate tables are required for each pitch and information is given which will never be used. A more compact presentation overcoming these problems is illustrated in table 2. Spans have been evaluated using equations 2, 3, and 4 for a UDL of 5kN/m² and pitch of 41mm, and the lowest (that is, the safe span) has been used.

To use the table we simply select the lightest section from those possessing a safe span equal to or greater than that required.

Table 2 Safe spans

Section (mm × mm)	Safe span (mm)	Weight (N/m²)
60 × 5	2918	577
55 × 5	2734	528
50 × 5	2545	480
45 × 5	2352	432
45 × 3	2070	259
40 × 5	2153	384
40 × 3	1861	230
35 × 5	1931	336
35 × 3	1628	201
30 × 5	1655	288
30 × 3	1396	173
25 × 5	1379	240
25 × 3	1163	144
20 × 5	1103	192
20 × 3	930	115

For example, at a span of 1850mm the 40 × 3mm section is the optimum. This was overlooked in both previous tables and could not be identified from figure **2** which would have resulted in 12.5% more material being used than necessary.

Alternative sections

Rectangular sections are stronger and more rigid when the depth is greater and sections which possess relatively high second moment of areas, such as I sections, are even better. They are, however, more expensive and so it is important to decide whether the material saving gained outweighs the cost disadvantage. Table 2 summarises optimum sections at various spans for different section sizes and configurations. Row 1 displays optimum sections for the company's initial product range and shows how 3mm-thick sections are used up to 2000mm span, followed by 5mm sections between 2250mm and 2750mm spans, with no sections suitable at 3000mm span. The effect of including 2mm- and 4mm-thick sections is shown in Row 2, where a material saving of 33% is possible up to 1000mm using 2mm breadths,

Table 3 Comparison of optimum sections

Design criterion		Clear span (mm)									
		750	1000	1250	1500	1750	2000	2250	2500	2750	3000
1 Rectangular section Initial product range Design to BS4592 b = 3 and 5mm only d/b ⩽ 15	d × b	20 × 3	25 × 3	30 × 3 3mm	35 × 3	40 × 3	45 × 3	45 × 5	50 × 5 5mm	60 × 5	
	weight	115	144	173	201	230	259	432	480	577	
2 Rectangular section Design to BS4592 b = 2, 3, 4 and 5mm d/b ⩽ 15	d × b	20 × 2 2mm	25 × 2	30 × 3	35 × 3 3mm	40 × 3	45 × 3	50 × 4	55 × 4 4mm	60 × 4	
	weight	77	96	173	201	230	259	384	423	462	
	material saving	33%	33%	0	0	0	0	11%	12%	20%	
3 Rectangular section Design to BS4592 b = 2, 3, 4 and 5mm d/b ⩽ 20	d × b	20 × 2	25 × 2 2mm	35 × 2	40 × 2	40 × 3	45 × 3 3mm	50 × 3	60 × 3	60 × 4 4mm	
	weight	77	96	134	153	230	259	288	346	462	
	material saving	33%	33%	22%	24%	0	0	33%	28%	20%	
4 I section Design to BS4592	d × b	20 × 2	25 × 2 2mm	30 × 2	30 × 3	35 × 3 3mm	45 × 3	45 × 4	50 × 4 4mm	55 × 4	60 × 5 5mm
	weight	77	96	115	173	201	259	346	389	423	577
	material saving	33%	33%	33%	14%	12%	0	20%	20%	27%	∞

and 4mm-thick sections entirely replace the 5mm ones for 2500mm and 2750mm.

Originally the depth/breadth ratio was limited to 15, but enquiries revealed that using an improved manufacturing fixture enabled aspect ratios of 20 to be achieved and tests upon panels did not indicate any buckling problems. Comparison of Row 3 with Row 2 shows the economies which may be gained using greater aspect ratios.

I sections may take many forms but, for comparative purposes, a section was chosen in which the flange was formed at the expense of the rib producing an aspect ratio of 20. This maintains identical unit weights to rectangular sections and the summary of optimum sections in Row 4 shows only a marginal improvement. Considering that an I section would be 30% more expensive than the same-weight rectangular section indicates that there is no net cost gain.

Further comparisons have been made under different deflection conditions, but these have been omitted for brevity.

Designing for bulk loads

Deflection is never a problem with concentrated loading applications, whether static or dynamic, so it is necessary to design to a permissible stress only. Loads may be applied at any position along a span and so designs are based upon the most adverse condition: that loads are distributed about the centre and supported by the fewest section bars in contact. The company's presentation was muddled, contained errors, and was not used by their own engineers. It is therefore not reproduced here. The objective though is clear: to present data as clearly and compactly as possible.

Bulk-loading theory

The forge-welded securing rods do not transmit any appreciable load between neighbouring support bars and therefore are not considered in the analysis. For a simply supported beam with a load which is partially distributed equally about the centre, as in figure **3**,

maximum bending moment,

$$M = \frac{W\,p}{4}(L - a/2)$$

equation 5

when $a = 0$, $M = \dfrac{W\,p\,L}{4}$

and when $a = \dfrac{L}{2}$, $M = \dfrac{W\,p\,L}{8}$ $\left(= \dfrac{w\,p\,L^2}{8}\right)$

for a rectangular section,

maximum stress, $\sigma = \dfrac{6M}{bd^2}$

equation 6

w = *Distributed load (n/m)*; W = *Point load (N)*

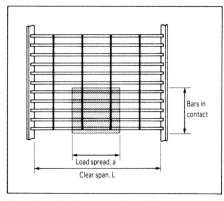

3 *Loading distribution under bulk loading*

Brochure presentation for bulk loads

The weight which a section can support will depend upon the nature of the loading and the contact area. For example a fork-lift truck which has solid wheels can exert a large dynamic pressure load, whereas storage items may produce smaller static loads over a wider contact area. These variations could be taken account of by using numerous safe-load tables, but because of the

Table 4 Safe concentrated loads (kN)

Bar section	Weight N/m²	Clear span (mm)					
		250	500	750	1000	1250	1500
60 × 5	577	6.00	3.00	2.00	1.50	1.20	1.00
55 × 5	528	5.04	2.52	1.68	1.26	1.00	.84
50 × 5	480	4.16	2.08	1.38	1.04	.83	.69
45 × 5	432	3.37	1.68	1.12	.84	.67	.56
45 × 3	259	2.02	1.01	.67	.50	.40	.33
40 × 5	384	2.66	1.33	.88	.66	.53	.44
40 × 3	230	1.60	.80	.53	.40	.32	.26
35 × 5	336	2.04	1.02	.68	.51	.40	.34
35 × 3	201	1.22	.61	.40	.30	.24	.20
30 × 5	288	1.50	.75	.50	.37	.30	.25
30 × 3	173	.90	.45	.30	.22	.18	.15

considerable number of possible load combinations this was thought undesirable.

Instead a more compact presentation was preferred which comprises a basic safe-load table and a system of factors to suit particular applications. Table 4 gives values of concentrated centre loads which single-bar sections can safely support and is produced using equation 5 with a permissible dynamic stress. If more than one bar supports the load, then the load carried will increase in direct proportion to the number of bars. If the load is spread along the span, then one of the factors may be used from table 5 which has been produced using equation 6.

To use the table to establish an economic section we use the following equation:

$$\text{equivalent load} = \frac{\text{maximum applied load}}{\text{number of bars} \times \text{spread factor} \times 1.3 \text{ if a static load}}$$

For example, if a dynamic 4kN load is to be carried by a span of 1500mm and the contact area is 150×150mm then, assuming a pitch of 25mm for load-bearing flooring, we find at least 6 bars are in contact. The calculated equivalent load becomes 635kN and from table 3, at a span of 1500mm, we see that the 40×5mm section is the most economic one with sufficient load-bearing capacity.

Table 5 Loading spread factors

Spread of loading relative to span	Point load	10%	20%	30%	40%	50%	60%	70%	80%	90%	UDL
Conversion factor	1.0	1.05	1.11	1.17	1.25	1.33	1.43	1.54	1.67	1.82	2.0

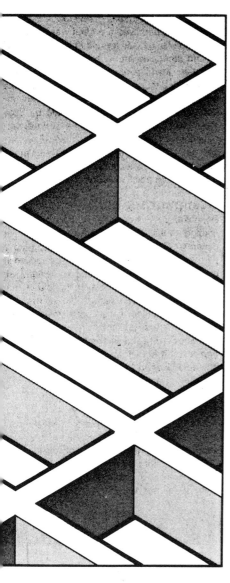

Discussion of findings

The ideas described have formed the basis of recommendations which were submitted to the company and have been implemented in the product range and a new brochure. Significant improvements have been made in displaying technical data and considerable material savings have been achieved. In addition it was suggested that the product range be based upon standard spacing gaps rather than pitches since the requirement is for a maximum gap. This favourably affects the economy of the thicker sections, particularly 5mm which more than satisfies the BS4592 requirement at 41mm pitch. This means that 25 bars may be used per metre width instead of 26; thus resulting in a 4% material saving and marginally reduced handling.

Application of educational projects

The content of this study has proved to be suitable for CAD analysis exercises and for enhancing the teaching of design optimisation theory. Because it is a simple product to describe, students studying mechanical element optimisation can readily identify the different design equations and the influence of functional requirement, geometric description, and material properties. Having a standard to adhere to draws students' attention to the importance of designing to imposed constraints and a recognised code of practice. Educationally projects can involve students in writing simple interactive programs which identify optimum sections under specified conditions or is producing complete tables and then using them to solve a problem. Indeed the potential for educational projects is quite considerable.

Reference
[1]Johnson, R C, *Optimum Design of Mechanical Elements*, Wiley.

Case Study by Mark Mitchell

Design process

The success of a product in the market-place depends primarily on a good product being met by a receptive customer. The process of achieving the successful transition of a good product concept into a physical product can be significantly affected by two vital factors: product requirement and design specification. These two factors occur at an early stage in product development and require careful analysis and communication. This article is the first in a two-part account examining the topics of product requirement and design specification.

In the following article general aspects and issues involved in product requirement and design specification are described. These aspects and issues are applied, in the next chapter, to a practical example: the product requirement/design specification stage of a domestic consumer product.

Definitions

To avoid confusion definitions and context should be made clear.

Product requirements, including the prerequisite of a good product concept, are the desired qualities and functions which the end-product must incorporate, or be capable of, to be successful in the market. These requirements, plus the essential factors of cost and time, are stated at the outset of the project by one of a number of interested parties. They may be stated directly by the product user or customer, but are usually stated by a manufacturer's marketing, technical, production, and distribution departments. The form that product requirements may take can vary. From the simplest case: a verbal discussion over a lunch meeting; to the complex: numerous experts sitting in committees to formally document many different considerations.

Design specifications are extracted from product requirements. A specification item could be described as stating in detail the design considerations derived from a product requirement: converting a basic idea into a more tangible, measurable form. Frequently the designer or design department will perform the task of translating product requirements into a workable set of specifications. If the product is uncomplicated then one designer could complete the translation task. Complex products, on the other hand, might require several designers or teams and related experts to produce a set of specifications.

The best way to describe the two items is by taking a basic example such as a grass cutting appliance. Marketing might say that a requirement is that 'the product must be capable of cutting long, coarse meadow-grass and short, fine lawn grass'. The subsequent specification might well state that 'long, coarse meadow-grass consists of such and such a type of grass, plus other plants, measuring up to 350mm in height'. An equivalent specification would be written to describe lawn grass. Later, at the design stage, suitable devices would be considered to achieve the required effect, consistent with parameters of cost, safety, and practicability.

It is helpful, at this stage, to establish three important definitions:

The designer is any person actively involved in the design of the product.
The user is a general term referring to the actual user or operator of the product.
The customer, in comparison, may not be the user. This is an important point. If the product involved is, for example, an industrial machine tool then the designer would not be designing the product only for the user. The customer's requirements would be a high priority; productivity, cost, and reliability being other obvious critical factors.

The user's or operator's needs would be important, but, as may be the case with some poorly considered products, certain requirements might be compromised in favour of other interests. This is an area of considerable debate. Product requirements and design specifications need careful preparation to produce balanced design solutions.

Products which are not balanced or well prepared run the risk of bias, mediocrity, and dismissive criticism.

Distinguishing differences

Distinctions between product requirements and design specifications should be recognised. There is a need to separate a sometimes vague or general notion from the analysis of that idea, in effect working from the general to the particular. Product requirements represent the expectations of those persons originating the product idea. If expectations cannot be met, or fail to be met, then the product will clearly cause disappointment and worse, rejection.

To avoid this, expectations or requirements should be converted into accurate specification statements, providing factual reference information to describe clearly the limits involved and the nature of the product. Equipped with the design specification the design team, or individual designer, should be in no doubt as to product design objectives. A side issue occurs here: in order to appreciate fully the implications of the design specification, the designer needs to understand the general context or background of the product. There is a risk in stating product requirements and design specifications too clinically. The designer may as a result produce an unsympathetic design solution and fail to appreciate the more subtle aspects of the product requirements. Thus the designer must be provided with, or personally research, the background to the product – not just bare facts but the general picture too. Design specification should be referenced to product requirement and the overall product context.

On this basis and with the design specification completed, the next stage, design work itself, can proceed. Design work may be carried out by an individual or a team, but always in conjunction with the individuals who produced the final specification, so that the general picture is not lost.

Real-world issues

So far this account has been based on an ideal form or textbook case. In the real world there is often insufficient time and sometimes even disinclination to work out all the details. Industry exerts considerable and sometimes overwhelming working pressure. It is not unknown for a manager or supplier of product requirements to assemble very minimal information. Besides pressures of work and time, individuals sometimes unprofessionally provide only vague, improvised requirements. Equally it is no longer possible to rely on intuition to provide information: facts are required. The specification writers, the design department, must either rectify unfair requirement deficiencies or, worse, proceed on the basis of assumptions. The result is invariably a mediocre, unsatisfactory product. The case for an impeccable product requirement and design specification must be recognised.

Planning and control

Staying in the real world, there is a need to underline planning and control. Essentially the subject of this article is planning and control. As mentioned earlier, time and cost are vital factors in preparing a product for the market. Every requirement must be analysed in the light of cost, time, and justification. For the product to be viable it must be planned from the outset, including market launch, duration of time in the market, redevelopment, and replacement. Good product ideas have failed through being too early or too late in the market-place. Planning involves anticipating changes and dealing with risk. Both potential problems can be accommodated by close analysis and contingency planning. Planning must be integrated into product requirements and thus design specifications.

Closely related to the subject of planning is the need to apply control. Designers are notorious for never being personally satisfied with the product of their work and continually searching for something better. Where non-designers would accept a proposed design at the point of reaching the required specification level, a designer may tend to pursue perfection. Unnecessarily high performance and increased costs could meet with disinterest in the market. Clearly controls must be applied to avoid this. Strictly applied design specifications enable checks to be made.

Related to this aspect of control and product requirement is the choice between existing mechanisms and standard components or fresh possibilities and innovations. Cost, time, resources, market expectations, and competition are the controlling factors in this case. Utilising existing mechanisms can lead to non-development, but equally innovation needs to be carefully managed to avoid runaway costs. Conversely innovation must be considered to a greater or lesser degree so that a company remains competitive.

Another aspect of control concerns the need continually to monitor any changes in the market likely to affect product requirements. Sometimes the market is liable to change very quickly – after the 1973 oil crisis for example. Here rapid changes had to be made as regards material processing and product economics. Again leisure products are especially prone to changes in fashion and product specifications must be constructed to respond accordingly. In the realm of design method textbooks, this process of information communication is termed feedback. Ideally feedback should take place continually and be equally responsive to inform product managers and, where practicable, permit changes. In practice the feedback mechanism has to be incorporated or planned at product requirement stage. The feedback lines which look so simple in the design methods diagram can prove troublesome to implement otherwise.

Valid information

Product requirements and design specifications both rely on information. The availability of information varies. It may be close at hand; it may be outside the organisation and remote. The latter type of information always carries the risk that it will be unintentionally overlooked. Alternatively if the requirement carries little risk and time is pressing, then there might be a temptation not to pursue difficult-to-obtain information. Individual organisations differ as to how information is collected, stored, and utilised. Whatever system is employed the producers of product requirements, and particularly design specifications, need relevant, accurate, and fully up-to-date material. The increase in product liability legislation, safety standards, and patent protection means that reliable and rapid information sources are even more crucial.

Increasingly electronic computer systems are being used to handle the competitive organisation's information needs, especially where design is concerned. In some industries standards and component development change frequently during the year, demanding new methods of coping with change. A critical if not sceptical attitude is needed in compiling product requirements and defining design specifications to check validity of information. For example, businesses involved in exporting products must be fully aware of a relevant country's product legislation and patent protection schemes. In addition, medical equipment manufacturers must know about safety standards and consider measures to prevent product misuse or malfunction. Any mishap in this area could prove fatal.

Product certification

Related to the subjects of information, product requirement, and design specification is the matter of product certification. Many established industries over the years have formed representative associations, with approval or certification schemes for members' products. This ensures that a standard is maintained and the customer has a guarantee of product integrity. With the increasing rate of technological development manufacturers are seizing the advantage to introduce innovations. The result is an expanding rate of new products. In some cases standards and certification authorities do not possess a direct precedent. Approval might consequently be delayed. To avoid this, and ensure acceptance to coincide with market launches of products, early consultation during the product requirement stage is vital.

Conclusion

The issues covered in this article regarding product requirement and design specification should not be considered definitive.

The aim has been to draw attention to these two stages in the product design process. It is hoped that the points mentioned will prompt further investigation and study. There is literature available on the subject, mainly referred to as a greater or lesser part of engineering design or the manufacturing process.

Case Study by Mark Mitchell

Product requirements and design specification

Product requirements and design specifications were introduced in the previous chapter. Here the topic is applied to a product design: the development of a domestic food processor. The purpose of the article is to describe the practical aspects of how product requirements are formed, how a design specification is subsequently derived, and where further information for the ensuing design stage is sought.

The article is fictitious in that it is not based on an actual case study. Marketing aspects can be represented in a comparatively realistic form and a marketing requirement is examined to illustrate the preparation of a design specification.

To review the topic briefly: product requirements and design specifications are seen as separate but closely related activities taking place after the product planning stage. The cycle of product supply is seen as product concept, planning, requirements/specification, design, development, manufacture, testing, and distribution stages.

Product requirements are the expected functions and qualities that the end-product must incorporate to be successful in the market-place. Design specifications are derived from product requirements and describe the necessary implications and factors to be considered in each requirement.

The specification, in effect, interprets the material meaning of the requirements indicating to the designer the parameters and opportunities within which to propose a design solution. The entire process relies for its success on a good product idea or viable concept, including the vital balance of cost and time factors.

Product ideas come from many sources in practice although ideas tend to be generated by two main influences. In some cases the marketing function supplies concepts identified in consumer requirement analysis. In other cases the technical function identifies opportunities in technological developments. The product idea featured in this article

originated from a food processor user's observations.

The essential purpose of the requirements/specification process is to communicate and convert a good idea into an effective material product – whether the scale of the product is that of an oil refinery, an aeroplane, or a software program.

Domestic food processor

The kitchen food processor is an appliance for rapidly converting basic foodstuffs into a form suitable for immediate cooking or consumption. For example; raw meat can be minced, fresh vegetables sliced, and dough and cake mixtures produced, with other operations possible according to foodstuff and processing-tool combinations.

The appliance has a circular processing bowl in which rotary tools are powered by an electric motor. The tools are detachable, enabling different functions to take place such as cutting, mixing, or shredding, according to tool-type. Product designs vary depending on motor-type, control methods, and level of utility or sophistication.

The product is associated mainly with France but manufacturers in the United States, West Germany, and Japan produce variants of two basic configurations. Once the novelty product was recognised as a useful and substantial item of equipment, food processors began appearing in the United Kingdom and the market is now led by a British-designed and manufactured product.

The food processor should not be confused with other food preparation equipment such as mixers, blenders, and liquidisers – these appliances tend to have individual functions. The advantage of the food processor is that it offers a wider range of functions; it is very fast in operation and the cost/function ratio represents better value than the other product categories.

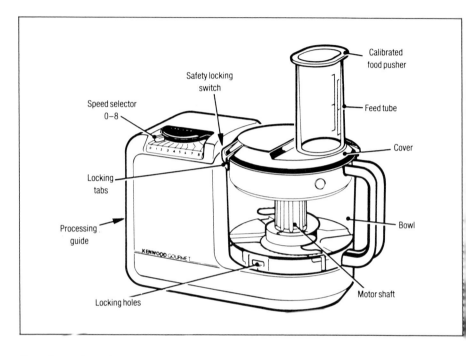

Speed selector
0-8

Safety locking
switch

Calibrated
food pusher

Feed tube

Cover

Locking
tabs

Processing
guide

KENWOOD GOURMET

Bowl

Locking holes

Motor shaft

Product development

The current generation of food processors suffers from many of the disadvantages common to portable products. They are bulky, cumbersome, trail power cables, are time-consuming to clean properly, and there are inevitable problems in storing the appliance and its accessories. To reduce these problems, a built-in food processor was suggested as a development.

The concept involves installing the electric motor and transmission beneath the kitchen work-top with a power take-off point recessed into the work-top surface with suitable safety precautions. Added facilities, such as the ability to attach other types of equipment (coffee mill, fruit-juice extractor, pasta maker, or can-openers), would be possible. Other advantages of the proposal would be avoiding the risk of dropping and possibly damaging the heavy

power-base, and reducing the risk of accidentally splashing liquid into appliance cooling apertures or controls.

Disadvantages are seen: if a built-in product fails, repair or replacement can be expensive. Built-in products also generally require a long-term commitment by the user. Frequent changes due to the whims of fashion are less practical.

Built-in products need special holes cut in the work-top and spaces provided in supporting structures. By comparison, portable appliances are usually easier to dispose of; built-in appliances must be disconnected and removed by skilled, qualified people.

For the purposes of this exercise a built-in approach was pursued. The concept of a built-in food preparation/processor is not new (there are few original ideas); indeed a similar product, manufactured in the United States, is available in Britain. This article was written independently, with an awareness of the American appliance.

Product planning

The design of a new product involves organising a senior management committee. Typically a group comprises managers from marketing, research and development, design, manufacturing, and sales: each representing the particular interests of their departments.

The design or technical manager may be appointed to 'drive' the project; in this example the design manager has been selected. At the outset overall plans are established including product concept, policy objectives, time-scales, and cost factors. Then the business of drawing up a list of product requirements commences.

Product requirements

Each participant in the project is requested to contribute representative and explicit requirements. These should be quantitative – how big, how small – and qualitative – attributes, standards, and characteristics of the product envisaged. Inevitably objective and subjective values enter into the formulation of requirements. This is why there is a need to request quantitative 'facts and figures' information; other less concrete matters can then be examined separately.

Product requirements can be said to exist in two forms, imperatives and 'bonuses' or, as described by Pahl and Beitz,[1] demands and wishes. Demands represent the vital, imperative factors that the product must incorporate. Failure to provide these factors means the product may not achieve market success.

Wishes, on the other hand, are requirements without which the product will still survive adequately in the market-place but which do in themselves enhance the product. Incorporation of wishes would of course add value or provide a bonus

quality to the product's performance, but may add unnecessary and excessive features, cost, and time to the project.

Once the requirements list has been drawn up, the design manager must next establish a requirements priority. Some demands may be more important than others and there are usually critical demands which must be dealt with first. The design manager must negotiate between conflicting interests.

Some matters may be dealt with in a meeting of two or three individuals; some may require a full committee meeting; others might have to be decided at a higher level of management. There may also be some requirements which exceed the organisation's resources.

Essentially the design manager must be a good listener, analyst, and communicator: a type of information broker, able to recognise different interests (ranging from styling to technical interests), and be practical, *and* maintain objectives.

The negotiations might take the form of consensus, compromise, or majority rule, depending on the organisation's character or operational bias. As mentioned in the last article, in the final analysis the product's success is decided by the customer.

If the product reflects a detrimental bias then the customer may not purchase. After the negotiating stage, the list of requirements is circulated to each interested individual or group, including those outside the project. This ensures that everyone concerned is fully aware of commitments. The list should not be considered unalterable; flexibility must exist to accommodate changes, either within or outside the organisation, such as problems revealed at the later specification stage or influential developments in the market-place.

Design specification

Once agreement is reached on product requirements, the design specification stage can proceed. Specially printed documents or

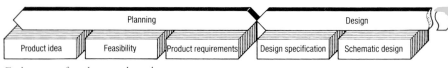

Early stages of product supply cycle

a computer program help in maintaining a standardised, recognisable format. A useful aid to writing the specifications and sticking to requirement objectives is to include the source-requirement as a reference on the specification form. This is suggested by Pahl and Beitz.

The specification must particularly allow for changes to be recorded and authorised. Often there is this provision, but pressures of work and tedious entry methods mean that changes are frequently only partially or sometimes never recorded. Many products fail because of poor documentation, indistinct recollection of events, and ill-informed decisions.

The specification change and authorisation method must be rapid and achieved with the least effort: an advantage of computerised documentation. The benefits of good specification documentation are predictably a record of context – why a change was necessary, a record of decisions and authorisation, a maintenance of vital objectives, adherence to standards, and a means of reviewing progress.

Process application

Applying the product requirement and design specification process to the food processor example, the marketing requirements would include the following considerations:

- function – foodstuffs conversion
- cost – value to the consumer
- prospective customer/user – socio-economic characteristics
- installation – qualified trade/DIY

- distribution – wholesale/retail
- image – United Kingdom/export
- competitors – performance/cost
- packaging – point-of-sale
- promotions – discounts/special product versions
- standards – British Standards Institution and representative consumer/trade groups
- legislation – acting/pending
- approvals – trade authority approvals/ recognition
- export – standards/legislation/patents/ trade factors
- obsolescence – product life-span
- safety – legislated/in-house standards/ testing

The marketing requirements relate directly and indirectly to technical, design, manufacturing, and user interests. To form the complete design specification, requirements from each department must be examined in detail. Analysis of individual requirements should reveal the many components making up the single requirement.

As this stage is reached, the task becomes a complex of interrelated factors. For example, the relationship of the marketing requirements to those of manufacturing must be identified.

It is here that a method approach is useful, particularly in large-scale projects. A suitable method reduces the risk of overlooking details or duplicating actions. The point is evident when more than two requirement-interests, or their components, relate. Often the factors involved are relevant to several different activities in the organisation.

176

Requirement example

Marketing's safety requirement, listed previously, provides a good example by which to describe the analysis task at work. Safety can be associated with the following factors, or requirement components, in a food processor design:

- electric power and transmission
- detachable tools and tool-holders eg cutting/mixing tools, can-opening attachment
- rotating parts, eg tools, power transmission
- storage of cutting tools
- protection of controls
- food hygiene.

Each of these factors or components of the safety requirement signifies an area for investigation. The design manager, or specification writer, must consider the implications of each component. For example, the last component, food hygiene, involves the following detail considerations:

- what type of foodstuffs will be prepared and processed?
- what are the extreme foodstuffs likely to be used?
- what are the particular constituents?
- what are the known characteristics?
- would there be reaction between these extremes of foodstuffs and engineering materials?
- what are the health hazards?
- what are the alternatives?

This list could be extended and the examination would eventually reveal patterns. These questions or principles should indicate to the specification writer where to seek answers to the questions, or where to seek further information.

An example is the question of toxic reaction between a foodstuff and an engineering material. The objective in this case is to establish what type of engineering material would be suitable. The question is vital if the manufacturing plant can use only particular materials.

In the question posed, toxic risk, there would be several avenues open for exploration; the foodstuff manufacturers, the representative trade association, government authorities, and the British Standards Institution would be possible sources of information on known materials. The engineering material supplier would also offer technical information and possibly a technical data supplier could be consulted. Test procedures would be included in these types of information and could be incorporated in the design specification.

In the case of known factors, information is generally available; problems and risks occur where unknown factors are encountered.

In the case of the food processor this problem of risk is unlikely. Suppliers of high-quality engineering materials or foodstuffs would test new products thoroughly before marketing them. It is when analysis reveals an unknown or obscure requirement factor, where information is limited or insubstantial, that specifications become difficult to construct.

In this case, as mentioned in the product requirement section, it is important to provide communication back to the requirement source. Either the requirement is modified or the specification stage is clarified: resources will be required to enable further investigations to be made – establishing research facilities or allocating time for an international information search, if necessary.

Specification summary

As described in the toxic risk example, each requirement may consist of components and these components may be further reduced to factors involved, and questions about the factors. Thus the requirement of safety involves the component, food hygiene. The component food hygiene involves factors such as toxic risk. Questions about toxic risk enable facts to be established to form the basis of a design specification – for example, what type of engineering materials do not react with conventional foodstuffs?

Applying this analytical process to each product requirement may seem unduly laborious, but this depends on methods and resources. Essentially the discipline of product requirement and design specification is an investment. The benefits are a thorough knowledge of the product and its various aspects. These include the product's market and end-user, its design, manufacture, and distribution.

The principle objective is to improve competitive edge within the market. It is always possible that a unique and otherwise unconsidered feature may be the result of a thorough requirement/specification process, thus lifting the product from the merely good to exceptional status.

Reference

[1] Pahl, G and Beitz, W, *Engineering Design*, The Design Council, London, 1984.

Clothes line

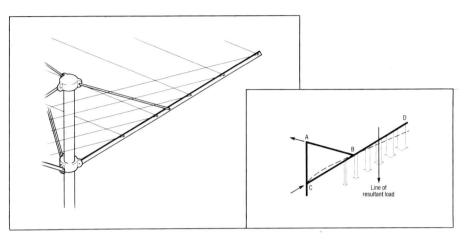

1 *Rotary clothes line*   **2** *Simple 2D force diagram*

In many gardens and yards the age-old single line for drying washing has been replaced by a folding structure on a single pole, with a spiral of clothes line providing the equivalent of about 20m of straight line. The original design work is not available, but a comparison of older models with ones now in the shops shows how an apparent lack of analysis in the early days has led to changes in even such an apparently simple structure as this.

Figure **1** shows the basic components of an early model. The central pole is thick and heavy. The main line support members are hollow tubes and the upper supports are simple bars bent at the ends to make the connections. All joints are pinned.

A simple diagram of one frame is shown in figure **2**. The loads on the main member get bigger towards the outer end giving a resultant load line well outside the joint B. The force in the upper support is a high tensile force in the order of three to four times the weight of the wet washing on this third of the structure.

Main member CD experiences both a high compressive load and a high bending load. At the supports, A and C , the horizontal

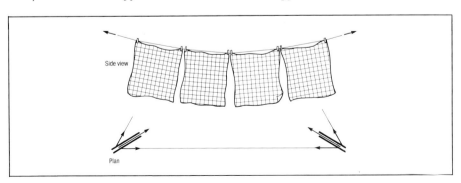

3 *Typical line*

loads are balanced by the other frames leaving the weight distributed between A and C.

This analysis is probably close to the original thinking since the members are designed to take just this sort of force system. In particular the upper member is a simple tie bar and definitely not designed for compression loads. Even under quite high winds one would expect any side loads would never reverse the effect of the high tensile forces just identified.

However, if a model of this type that has been in service is examined, almost invariably these upper members are bent and if you have seen one collapse you will know that one of the main members rotates upwards, often in a quite moderate breeze or even no breeze at all. What is wrong with our analysis and why are our expectations not realised?

Let us first look at one of the lengths of line, figure **3**. The weight of the washing is supported by tensile forces in the line. These will act at an angle near to the horizontal. In addition a plan view shows that the resultant force from adjacent lines will give load radially inwards at each frame. Thus we need to redraw our force diagram as shown in figure **4**.

Main member, CD, is still carrying compression and bending, but now the bending is in the opposite direction. The big change is in member AB which now carries a large compression load. In addition support A now sees a total upward load and support C alone carries more load than just the weight of washing. Therefore it is not surprising that AB frequently buckles.

More recent models of this device have used tubes or channels for the upper members which are much better at

withstanding compression loads. The choice frequently depends on the way in which the frames fold.

Tubes have to be offset to avoid fouling the main member; channels can be made to nest round the main tube when folded. If a full analysis had been made for the original design, this change would not have been necessary and a greater level of customer satisfaction would have been achieved.

A study of the reasons for a failure often reveals omissions in the design process. More often than not some factor has not been given sufficient priority or consideration to reveal the real nature of the problem as apparently happened here.

Such omissions happen on the best-run projects and show the necessity of having a good checking and review procedure for all parts of the design. No product should be considered as too trivial to be checked. Many hours of customer frustration probably cost many thousands of pounds in lost sales and repeat orders in this use alone.

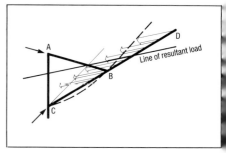

4 *Corrected force diagram*

Design Analyses

Tolerances

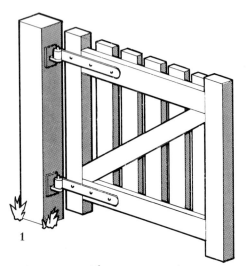

1

If it were possible to measure and manufacture physical objects with absolute precision at a reasonable price the design task, and indeed many everyday problems, would be much simpler. If you could measure the walls of a room and so decide exactly what lengths of wallpaper you needed to cover them, your local shop could supply them ready-cut and your decorating task would be much easier. If you could mix exactly the right amount of concrete to cover your garden path, you would not have an extra bucketful over at the end to get rid of. If only . . . but it is not possible. The history of technical advance can be described in terms of an increasing ability to measure and manufacture to smaller and smaller tolerances at an acceptable price, and of finding ways of accomplishing tasks without needing impossibly small tolerances.

But why bother, apart from special instances like screw threads or mating faces, why worry about a millimetre or two? Take a simple example. Consider a garden gate hanging on two pin hinges (figure 1). The two hinge pins are mounted one above the other on the gate post and two looped straps are similarly mounted on the gate. The fit of the pins in the loops is loose enough not to

cause alignment problems and the gate will swing well, but how is the weight of the gate carried? Each pin has a shouldered base on which the strap loop may rest. However if the distance between the loops on the gate is greater than the distance between the pin shoulders, the top loop will not touch its shoulder and all the weight of the gate will be carried by the bottom hinge (figure 2). Conversely if the distance between the loops is less than that between the pins, all of the weight will rest on the top hinge (figure 3). Each may be preferable under different circumstances, but if you want one of them you must limit the freedom of the manufacturer to meet the dimensional requirements. If you wish to approach even the mid-position where the load is shared by making the two dimensions virtually equal then the cost of achieving it would exceed the rest of the cost of the gate.

There is a way out. If one pin is lengthened and lowered to accommodate a spring of appropriate stiffness, the proportion of load sharing between the hinges becomes far less sensitive to the variation in distance. Hence a reasonable share of load can be achieved without excessive cost (figure 4).

It is usual to think of tolerances being applied to lengths, but almost any measurable quantity may have a tolerance applied to it,

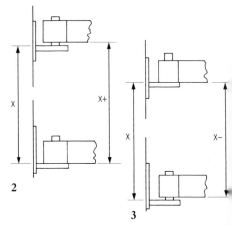

2

3

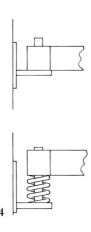

4

even when the measure is partly subjective. Thus a standard range of colours of paint can have specified tolerances covering the narrow band of colours which can all be considered as 'cobalt blue' for example.

The tolerances specified by a designer may be called up directly, as an allowable variation in a manufactured length, or indirectly by specifying a British Standard, or some other standard or code of practice. Variations in measurement and manufacture happen to large dimensions as well as small ones. The designer must balance the costs of achieving smaller tolerances against the costs which may have to be borne if parts do not fit or perform correctly because of poorly specified tolerances. Achieving a required tolerance is a two-part process. The appropriate method of manufacture must be chosen, set up, and checked periodically during the production run. The parts that are produced must be inspected, either on a sample basis or item by item, to ensure that they are acceptable and corrective action must be taken if they are not.

Tolerances vary with the type of industry. Thus the main piers of the Thames Barrage had to have their critical sections positioned to within 5mm to accept the movable gates when they were delivered. In civil engineering such measurements must be made by surveying instruments since it is impossible to use any form of physical measuring device in the middle of a fast-flowing tidal estuary.

In the aerospace industry, where weight is critical, the confidence which can be placed in the strength of lightweight structures is dependent on the accuracy of manufacture. Hence the tolerance requirements are of a much higher order. When Grumman were preparing to manufacture parts for the Apollo programme, it was necessary to position accurately two vertical planes about 20 metres apart. It was soon realised that 'vertical' was a relative term and the curvature of the earth had to be allowed for to achieve the required tolerance.

Such requirements seem almost rough when compared with those for some scientific instruments. The primary mirror, manufactured by Perkin-Elmer for the Space Telescope launched in 1985 has one of the most accurate reflecting surfaces ever produced. Over a diameter of 2.4m (94in) the surface varies from its ideal concave hyperboloid shape by less than 0.04 microns (10^{-6}m). If this mirror were scaled up to cover the whole of Europe, some 4500km (3000 miles) in diameter, the maximum variation from the ideal surface would still only be about 6cm ($2\frac{1}{2}$in). It took 20 months of grinding and a further eight months of polishing to achieve this accuracy. Measurements were made by a laser interferometer.

Putting tolerance figures on drawings may just seem a tedious detail but they could make the difference between the design working or not, or being competitive, or overpriced.

Weight control

Of all the factors affecting engineering design, the one which has a significant effect on the performance, cost and marketability of most products is weight. This fact seems so obvious and weight is such a simple concept that it is frequently neglected while more complex factors are given prominence. In this context, weight and mass have a equivalence since gravitational attraction can rarely be adjusted to suit the designer.

Performance

Mass, of course, appears as a multiplier in many calculations of dynamic or structural performance. Thus the efficiency of many products varies at least linearly with the mass of its moving or structural components. The distribution of mass can have significant effects where rotational inertia or structural vibration are important factors.

Consider, in simple terms, the structure of a multi-storey building. Take a basic vertical load case and consider the top storey initially. It carries the weight of the roof plus rain and snow loads. If the structure is heavier than necessary then the next storey has to carry that excess weight as well as the roof loads, the top floor loads and the optimum weight of the top floor structure. If the structure of that storey is similarly overweight then the loads on the next storey are excessive and so on down the building. A factor of a few per cent in each case can increase the foundation loads, and hence their size and cost, by perhaps 15–20% on a ten-storey building and put 10% on the cost of the building.

A number of modern cars are fitted with air dams and spoilers to reduce the aerodynamic drag. They certainly work, as wind tunnel tests show, but they also add weight to the car. Thus at lower speeds where drag is less significant the performance of the car in terms of fuel economy is actually worse. Generally speaking such accessories are only beneficial if the car spends most of its travelling time at speeds above 50mph, a rare condition.

Apparently minor details can add up to significant savings. An American fighter plane was made 50kg lighter by using hyperbolic stress relief curves in machined corners rather than the usual fillet radius. With numerically controlled machining the extra cost was insignificant. Even the weight of paints should be examined carefully. The weight of paint on a single railway vehicle can be over a tonne.

Cost

The costs of materials even in a semi-finished state such as castings or extrusions are usually directly linked to weight. However the relationship to final costs is often more complex. Consider a beam to carry a load. Analysis may show that a standard steel section would adequately carry the loads at the most highly stressed section, but if used for the whole beam would be under-utilised elsewhere. It may be possible to cut lightening holes in the understressed areas. Alternatively a smaller section could be used and reinforced where necessary. Either way may save weight but would be more expensive. The trade-off between weight and cost then becomes a matter of judgment.

In some cases an actual figure can be put on the economic advantages of a reduction in weight in performance terms. This figure then prescribes the upper limit of worthwhile spending to achieve the reduction in $£$/kg. Sometimes a quantum factor appears. Thus a saving of 100kg on the weight of an aircraft might allow an extra passenger plus seat etc to be accommodated, whereas a saving of 120kg would only have the same effect and thus not be worth the extra cost. Useful savings may therefore only be made in steps. Occasionally these figures vary for different parts of the design. A small weight saving on the wheels and suspension of a car gives considerable benefits on the ride and it is worth spending a sum on this which would not be warranted to give a similar weight saving on the body. Higher spending limits usually apply to moving parts or those which are only needed occasionally such as car jacks, emergency escape gear or access doors.

Marketability

Improvements in performance or increased perceived value for money due to weight changes will also affect marketability, but in many cases the effect is more direct. Equipment which must be carried or man-handled will sell better if it is lighter even if it is more expensive. Larger items may make savings on the cost of supports and foundations, by being lighter, which exceed the extra cost of the equipment.

Another marketing advantage comes from the easing of transport problems when delivering lighter goods. Warehousing is also easier with lighter packages.

Weight is an apparently mundane factor in the design process and calculating the weight of drawn components can be a very tedious task. However the effects outlined above can make the difference between success and failure for an otherwise good design.

Design Perspective

Rope

A rope is a flexible bundle of fibres with tensile strength. The rope may be over 100m long but the fibres are rarely more than 1m long. The only exceptions to this are wire ropes, which are bundles of continuous wires, and some modern synthetic ropes where very long fibres are available. The only forces holding the fibres together are those friction forces generated by the method of construction. In this context the term 'rope' covers all sizes from sewing thread to ships' anchor cables.

A knot is an interwoven construction of rope intended to tie a rope to itself, to another rope, or to another object. A knot should serve a prescribed purpose, either decorative or practical, and should hold its form when tied. This broad field covers the range from simple thumb knots to complex weaves. The range can be divided into hitches, bends, knobs, loops, splices, and sinnets. A hitch ties a rope to another object,

a bend ties two rope-ends together, and a knob is tied in a rope to locally enlarge it. A loop knot ties a rope back to itself, a splice is a knot formed by tying the separate strands of a rope, and a sinnet is a rope formed by plaiting or weaving strands together rather than twisting them.

Construction

A rope is constructed from fibres by a simple twisting process repeated in stages. If a quantity of fibres is carded so that fibres lie parallel to each other and a small group of fibres is pulled steadily from the rest and simultaneously twisted, a surprisingly uniform 'yarn' is formed. The way in which this happens requires some explanation. The steady pull causes the fibres to slide over each other but the twist tends to press them together and the resulting friction resists this slide. Where the yarn is thinner, with fewer fibres, it twists more and hence resists being made thinner still. Thicker parts of the yarn have little twist and the fibres there slide apart until the thickness matches the rest of the yarn. The thickest part of all is the original fibre bundle where more fibres begin to twist and slide out to continue forming the yarn. This process was first carried out by hand, then with the aid of bobbins and spinning wheels, and now entirely automatically, but it still remains essentially the same process.

Yarns normally are given a right-handed twist. For the next stage one end of a bundle of yarns is held and the individual yarns twisted further to the right. They are then allowed to untwist as a bundle and thus wrap themselves tightly round each other to form a left-handed 'strand'. This process is called 'laying up'. If only two or three yarns are twisted together a 'thread' is formed. This method naturally produces a left-handed twist. Right-handed strands are sometimes produced by laying up a series of threads. This process is then repeated to twist the strands into a 'rope'. Until recently this

was done by stringing out yarns down the length of a 'rope walk' and carrying out all the twisting from one end. The length of rope which could be made in one piece was limited by the length of the walk. Modern machinery can now produce rope continuously and lengths are only limited by handling and storage requirements.

If the process were taken a stage further, ropes would be laid up to form cables and once more, the cables would twist to form a hawser. The twists would reverse at each stage. Right-handed rope is often called 'hawser-laid' and left-handed rope 'cable-laid'.

The deceptively simple method of producing a twisted rope does in fact produce a very complex structure where even the geometry of a single fibre is difficult to describe. This structure of man's oldest product has probably never been fully analysed and would certainly provide a research task worthy of a PhD but admittedly of little immediate practical application. Ropes of different materials are easily made and tested so a rigorous analysis is not necessary. However, similar research has in the past led to new insights and developments and such may well be the case here one day.

History

The method of rope manufacture is one of the earliest production processes, invented by man probably at least 20,000 years ago. Perhaps the earliest-known rope is a fossilised piece 30cm long and 7mm in diameter found in the Lascaux caves in France and dated at 15,000 BC.[1]

It was made from plant fibre and was a two-strand cable-laid rope indistinguishable from a modern equivalent. Ropes in this twisted form may well have been preceded by plant fibre tied into bundles and tied end to end to achieve the desired strength and length. Plant fibre, vines, and dried grasses would probably have been favoured over animal materials for rope because of its greater length and abundance;

although animal sinew and hide strips would have been better for bow strings, lashing axe heads, and similar uses.

For many thousands of years rope was a major structural element, fastening device, and mechanical linkage in a designer's kit of component parts. In its structural role it formed, and still forms, the major part of suspension bridges, the tensile members of the truss structure of a sailing ship, and almost all of any tent. As a fastener it lashed together primitive huts and the masts and spars of the ships which first circumnavigated the world. Mechanically ropes have served many purposes apart from the movement usually possible in suspended structures. Winch and pulley systems come most readily to mind and are still used in cranes, lifts, and many other devices. Drive belts are a development of specialised 'rope'. Twisted bundles of rope became torsion springs to power catapults and other war machines for the Roman armies.[2] The flexibility of a rope provides that physical manifestation of exponential growth in the

1 *Half hitch*

2 *Clove hitch*

crack of a whip, where the initial movement of a loop in a chord increases speed until the end breaks the sound barrier.

The story of the rope cannot be separated from the parallel story of the knot. The simplest knot is the half hitch, a single turn of rope round a bar and the end trapped under the rope (figure **1**): not a very safe knot by itself but sometimes useful. Two half hitches back to back form a clove hitch (figure **2**): useful for pulls perpendicular to a bar or for tying bundles tightly together.

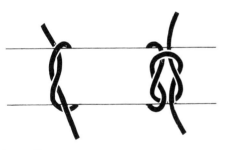

3 *Reef knot*

All of its crossings should lie on a convex surface. If the end of a half hitch is taken over the rope and tucked through again, it becomes a half or thumb knot well known by those who deny all knowledge of knots (figure 3). Its name derives from the method of tying where the thumb is usually used to tuck the end through the loop. In fact it is almost impossible to tie any kind of knot without using a thumb; indeed a twisted rope cannot be made by hand without an opposed thumb. It is tantalising to speculate that man was encouraged to develop the opposed thumb by the desire to grasp and manipulate vines and creepers to make knots and ropes.

The simple elaboration of the basic half hitch can be continued to produce a range of knots suitable for many purposes from structural lashings and making nets to tying up bundles for transport. These knots must have been known at a very early stage in man's history. Knots for harnesses for both ridden and pack animals must have expanded the range of both practical and decorative knots, but the variety of knots needed and the time to elaborate them increased with the era of the sailing ship. The simple square-rigged vessels of the Vikings and the Phoenicians carried only a few stays to support the mast and the necessary ropes to support and control the sail. After more than a thousand years of evolution in design the ships that fought the Napoleonic wars had become great complexes of rope and wood with iron making its appearance in appropriate places.

4 *Two half hitches*

5 *Buntline hitch*

6 *Midshipman hitch*

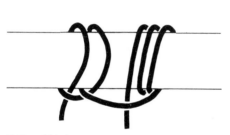

7 *Camel hitch*

8 *Bowline*

Nelson's *Victory*, for example, used miles of rope; some as 'standing rigging' forming part of the structure of the vessel giving it both the strength to withstand the wind loads on large areas of sail and the flexibility to absorb the buffeting of wind and waves and the shock loading from continuous cannon fire. *Victory* carried thirty 12-pounder cannon, twenty-eight 24-pounders and thirty 32-pounders. A 32-pounder weighed three tons and recoiled seven feet when fired. Most of the rest of the rope was 'running rigging': the complex control system of the ship allowing each sail to be set at the right angle to catch the wind.

The variety of knots in use, not only at sea, but in a number of other trades, is considerable. One of the largest published

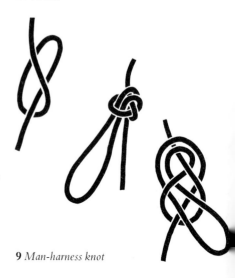

9 *Man-harness knot*

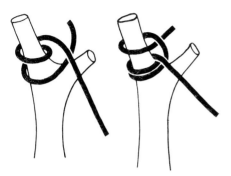

10 *Rolling hitch*

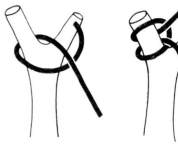

11 *Cow hitch*

12 *Granny knot*

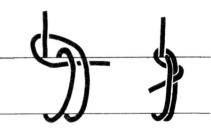

13 *Yard hitch*

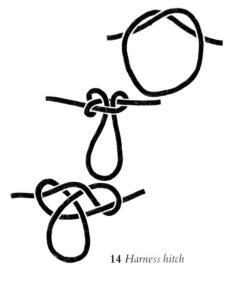

14 *Harness hitch*

collections[3] lists nearly 4000 and many of these are generic types which may each be expanded to make hundreds more. That collection does not include a number of knots specially developed for modern ropes of synthetic materials, which have low friction surfaces and may be more flexible than the natural fibre ropes. In addition new knots are still being invented (or rediscovered); some being deliberately designed for a specific purpose.

Allied to knots are the various cleats, frictions, grips, and cam and wedge devices for tying off or attaching ropes to convenient places. These bear the same relationship to knots as chains, belts, straps, webbing, and other flexible materials do to rope itself.

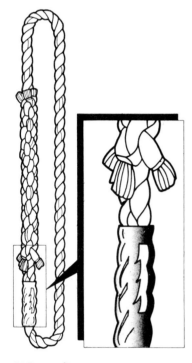

15 *Long splice*

Rope mechanisms

The mechanical devices made possible by rope and its unique properties are frequently neglected in engineering courses, but they do feature in so many designs that perhaps a little more emphasis should be placed on them. Consider some of the simplest rope mechanisms. A single pulley changes the direction of a rope and the load it is carrying. The diameter of the 'sheave' (the pulley wheel) should be chosen to match the size of the rope and the material it is made from since the rope will be damaged if bent to too tight a curve. The radius of the groove on the sheave should match the radius of the rope. Too narrow a groove will pinch and distort the rope and can cause jamming; too large a groove concentrates the side load on the rope and may flatten it and cause excessive wear. Pulleys should be mounted to avoid side loads on the sheave. Ships' pulleys were often themselves suspended on a rope to make them self-aligning.

A drive belt for power transmission is common in mechanical systems. An endless belt carrying buckets to raise water is described on pages 20, 21. To use the *Victory* as a source again, it had just such a bucket chain to pump out the bilges but it also had a much more powerful drive belt to raise the anchor. The anchors were connected to the ship by hawsers 24 inches in circumference (20cm diameter). These were far too large to go round the capstans which instead drove a continuous rope loop a mere 11 inches in circumference (9cm diameter) known as a messenger. It passed four times round the capstan drum then out in a long loop alongside the anchor hawser. The two ropes were then 'nipped' together, that is, tied to each other by short lengths of cord. 'Nipping boys' ran back and forth alongside the ropes, taking off the cords at the capstan end and tying them back on at the other end of the loop. The loop was so arranged that it could be used to 'weigh' either the port or starboard anchor depending on the direction of rotation of the capstan. Some idea of the loads carried by these ropes may be gained by realising that the capstan bars were manned by a crew of 280.

Wire ropes

These deserve a special mention since they are the ones most often used in engineering applications. Unlike fibre ropes, wire ropes should never be knotted, except for splicing

194

to join two ropes or make a loop. This should be done with care since it requires skill and special tools and can be dangerous if single wires break and whip. Most connections are made via clamps or swaged joints. These latter joints are made by putting the rope or ropes into a softer metal sleeve and applying pressure to squeeze the softer metal around the rope forming a solid block (figure **16**).

Wire rope has a reasonable capacity to resist longitudinal compression provided it is supported by a close-fitting tube to stop it buckling. This is the basis of a number of push-pull control cables used on everything from bicycle brakes to early aircraft controls. Special forms of wire rope are sometimes used to transmit torque in power transmissions. Torque can only be transmitted in the direction which tightens the lay of the rope. Its great advantage is flexibility which allows it to accommodate a certain amount of misalignment. However, at higher speeds the centrifugal loads cause the rope to 'whirl' like a skipping rope.

The most familiar use of wire rope is for raising loads on cranes and in lift shafts. Since this type of application has received a lot of attention it is well covered by various codes of practice and regulations. Factors of safety used for these purposes often seem very high, but stress concentrations at pulleys and fixing points can be extreme. In addition shock and dynamic loads and a long life expectancy make such factors necessary.

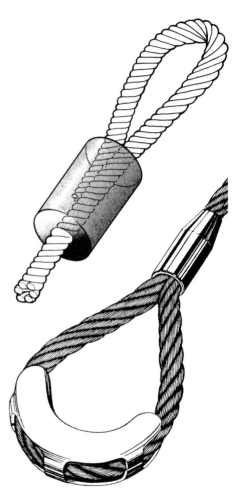

16 *Swaged joints*

References

[1]Lersi-Gourham, A, 'The archaeology of Lascaux cave', in *Scientific American*, June 1982.

[2]Soedel, W and Foley, V, 'Ancient catapults', in *Scientific American*, March 1979.

[3]Ashley, C W, *The Ashley Book of Knots*, Faber and Faber, 1947 (still in print). First published by Doubleday, New York, 1944.

Design Management

Critical path analysis: 1 networks

Critical path analysis, CPA, and the closely related system programme evaluation and review technique, PERT, grew out of methods evolved during World War II to deal with the organisational difficulties of the supply of military and civilian requirements under difficult conditions. These methods were developed in the United Kingdom by the Central Electricity Generating Board for overhauling power stations and in the late 1950s reduced the average time to one third of earlier comparable times. Thus CPA was developed where the times for individual tasks were well known but the organisation of the whole project, despite a long history, could be considerably improved. In the United States, PERT was developed by their Navy for the Polaris project. Here the basic problems were the unprecedented complexity of the project and the uncertainty of the times to complete the component tasks. A comparison of the actual programme with the original plans shows a saving of two years on the project. Despite the completely different type of problem, the two methods are very similar. A slightly different approach in France resulted in a method, very similar to CPA, known in the United Kingdom as precedence diagrams. The only real difference was in the presentation of the project sequences now known as an 'activity-on-node' network rather than the 'activity-on-arrow' network of CPA.

In the years since 1960 network techniques have become accepted as a reliable method of planning and controlling complex projects in a wide range of industries. Computer programs are available to help draw and analyse a network and there are two British Standards directly concerned with these activities.[1,2] The methods have also expanded to include time-related costs and resource allocation.

This teaching aid is intended to give an introduction to the initial stages of setting up a network and allocating times to tasks.

Later chapters will look at further planning techniques and project control. This is of necessity an outline and if any part is not clear, the references at the end should be consulted for a fuller explanation. Computers are not necessary for CPA for even quite extensive projects and can even prove a hindrance in the early stages. However, it should be noted that the basic principles are deceptively simple and care should be taken to perceive the complex messages which are the sum of the easy details.

Logic of a project

Many industries have preconceived ideas about how projects should be run and which are the important tasks. One of the earliest industries to take up CPA was the shipbuilding industry. For them the key event which marked the progress of a new ship was the launch. The early networks showed a maze of activities conducted on the slipway all culminating in the all-important launch; then a second phase which took place in the dock. Soon it was realised that many activities, such as the installation of electrical wiring, plumbing, air-conditioning, and other services and the work of carpenters and painters, did not in fact have to reach a specific point before the launch. Work simply stopped for the launch day then carried on in a different location. Soon the picture of the project had changed completely. The launch became a minor activity which could take place any time after the hull was watertight and painted. Other activities were more important and simply the discipline of examining the logic of building a ship gave greater understanding of the process than had been derived from thousands of years of history.

The logic of a project can be examined to various levels of detail depending on the reasons for the examination. There is a tendency for newcomers to go into too much detail so a sense of proportion must be retained. The first step in CPA is to draw a diagram or network which represents the logical sequence of individual activities which make up the project. Although there is a time sequence there should be no attempt to use a time-scale at this stage. It is useful to begin by drawing up a list of all the activities you can think of. Note that each activity should be an identifiable independent operation or process which takes time to complete and not a point in time which represents the completion of a stage of the project. On a CPA network each activity is represented by an arrow. The tail of the arrow is the start of the activity and the head is its completion.

The arrows are arranged in a network representing the logic of the project. Time conventionally flows from left to right. Each node on the network represents an event in time. This event cannot occur until all the activities (arrows) leading to it are complete. Similarly none of the activities leading from the event can begin until the event has occurred. It is soon necessary to introduce dummy activities to clarify the logic and avoid unnecessary constraints. If each activity on the list is examined by asking the questions, 'Which other activities must be completed before this one can begin?' and, 'Which other activities cannot begin until this one is complete?', the logic begins to appear.

There are two particularly important events on the network. The start event marks the beginning of the project and all activities which begin there require no preceding work. The finish event marks the end of the project and occurs when all activities leading to it are complete. There can only be one start and one finish event on a standard CPA network. Drawing time flow from left to right helps to eliminate accidental loops. Multiple start or finish networks and continuous processes require other techniques beyond the scope of this article. A few extra dummies perhaps representing time delays or date sequences

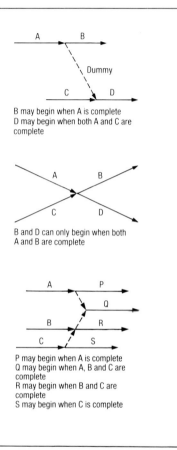

B may begin when A is complete
D may begin when both A and C are complete

B and D can only begin when both A and B are complete

P may begin when A is complete
Q may begin when A, B and C are complete
R may begin when B and C are complete
S may begin when C is complete

1 *Activity arrows*

can easily bring projects with multiple delivery dates or limited plant availability within the scope of CPA.

At the early stages you will probably redraw the network a number of times, so quick sketches are recommended before a final, neater version is drawn up. You should learn a lot about the project at this stage and find a number of new activities and logic links you had not thought of. It helps if several people are involved, including those responsible for the project planning, since the analysis will be useless if the network does not represent what will actually happen.

The network represents a statement of intent and may well be incorporated into a contract, so should carry the authority and agreement of the project manager. This does not imply that a network is inviolate since unforeseen circumstances may alter the project as it proceeds. However, the agreed completion of the network should be seen as an important step in the management of the project and any further changes should not be lightly regarded.

Exercise 1

Take a simple task you are familiar with, such as making a cup of tea or cooking a simple breakfast, and draw a network representing the logic of the activities. Note that an activity which consumes time like boiling a kettle or making toast in a toaster does not need an operator. Do not include any restrictions imposed by having only one pair of hands; the network shows the logic of the project at this stage, resource allocation comes later.

Exercise 2

Construct the network for this logic:
- activities A and Q may start at the beginning of the project;
- when activity A is complete, activities B, C, and D may start;
- when B is complete, E may start;
- F and G may start when C is complete;
- J can start when E and F are complete;
- when J is complete, L may start;
- M may start when L and G are complete;
- H depends for its start upon the completion of D;
- when G is complete, K may start;
- N may start when H and K are complete;
- P may be carried out when M, N and Q are all complete;
- when N and Q are complete, R may be started;

- S may start when P is complete;
- the start of Y depends upon the completion of both S and T. Y is a final activity;
- when P and R are complete, T, U and V may start;
- W can start only when both U and V are complete. W is a final activity;
- X may be started when T is complete. X is a final activity; and
- when S is finished, Z may start. Z is a final activity.

Take care: the information is not in a simple order and dummies will be needed. Check the logic carefully; this network will be used for a later exercise.

Adding the times

For the purpose of analysis, by hand or computer, it is useful to refer to the activities by numbers. It is convenient to number the nodes and refer to the activities by their tail and head numbers, sometimes called i–j numbers. An added advantage is that a simple list of activity numbers, including dummies, automatically gives the logic of the network, although the diagram generated from the numbers may not resemble the original. Numbers are usually allocated from the left such that the tail event of any activity is lower than its head event. Numbering is often done in steps of five or ten, similar to computer program listings, to allow new activities to be added with minimal renumbering. Each activity should have a unique number pair; however, you may find that two activities run in parallel between the same nodes and would therefore have the same numbers. Activities U and V in exercise 2 are such a case. An extra dummy is added to the beginning of one of the parallel paths to make the numbers unique. To put the numbers on the events by convention each node has a three-compartment circle. The event number goes in the left-hand compartment. The other two will be used later.

In CPA it is assumed that the time to complete each activity is known. If there is some uncertainty, a good statistical estimate can be obtained from the formula:

$$t = (a + 4m + b)/6$$

where a is the shortest possible expected time,

m is the most likely time,

and b is the longest possible expected time.

PERT is essentially an extension of this approach, giving probability figures for all calculated times throughout the following analysis. The necessity for this is rare and its usefulness is dubious unless the users are conversant with statistical arguments. PERT is also impractical without the use of a computer. CPA is frequently calculated manually even on projects with three or four hundred activities.

The time for each activity is written in on the appropriate arrow. Dummies are assumed to have zero duration. The analysis can now begin by calculating the earliest time when each event can occur The start event occurs at time zero. The events are then followed in numerical order and the earliest finish time for each of its preceding activities calculated by simply adding the activity duration to the earliest event time of their tail events. The latest of these finish times is then the earliest time when the event can occur. These earliest event times, EET's, are written in the upper of the two remaining compartments of the event symbol.

This process can be continued until all of the EETs have been calculated. The EET of the finish event is the earliest time in which the whole project can be completed. This procedure finds for each event the length, in time, of the longest path through the network leading to that event. Having found the total project duration we may assume that we want to finish the project as soon as possible and this time therefore also represents the latest event time, LET, of the final event. If we now go through the events in reverse order, this time

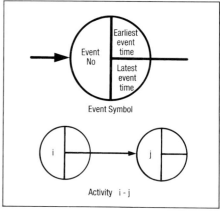

2 *Notation*

subtracting the times of subsequent activities from the LET of their head events we can select the time which is the latest that the event can occur without delaying the entire project. These LETs are written in the last compartment of the event circle. The LET of the start event should be zero. This acts as a partial check on the arithmetic.

If you examine the network you should see at least one chain of activities running through the network where the EETs and LETs are the same and the activities forming the chain have durations equal to the difference between their head and tail event times. This chain is the longest path in time

Exercise 3

Add the following times to your solution to exercise 2 and find the EETs, LETs, and the critical path.

A	4	F	7	L	3	R	11	W	2
B	2	G	6	M	6	S	7	X	3
C	3	H	5	N	8	T	4	Y	5
D	8	J	10	P	11	U	3	Z	3
E	1	K	5	Q	6	V	8		

through the network and determines the length of the project. This is called the critical path and its activities are marked with a double line to clarify the sequence.

Reducing project duration

This is the point where the initial consequences of the choice of project logic are apparent. If the project duration is too long, the critical path or paths must be shortened. Thus you may concentrate your attention where it matters to re-examine the logic of the critical path and the activity times allocated to check their validity. If reductions in the critical path times can be made, the rest of the network should then be checked since other activities may have become critical.

One of the main ways of shortening times comes from the realisation that in a particular sequence of tasks each one does not need to be completed before starting the next. Provided a certain minimum has been done the next activity can begin. Similarly when one activity is complete there will still be a necessary minimum of the next one to complete. Where there is no interaction with the rest of the network only these 'leading' and 'lagging' times are important to set up a ladder of activities. For example, consider the production and distribution of a series of diagrams as part of a larger project. Initially this may have been represented as a series of activities as shown in figure 3, but by changing them to a ladder a considerable saving can be made provided it is possible to organise the work in this way.

Next steps

Part 2 of this series examines the non-critical areas of the network to show how the resources available, such as staff and production facilities, may be allocated and how costing can be carried out. The third and final part examines project control using CPA.

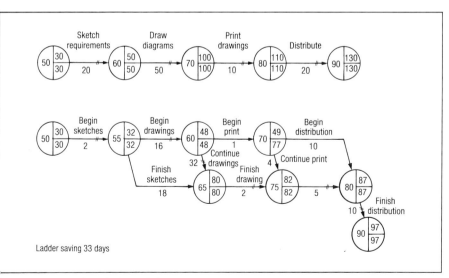

Ladder saving 33 days

3 *Two part networks. The lower ladder structure saves time*

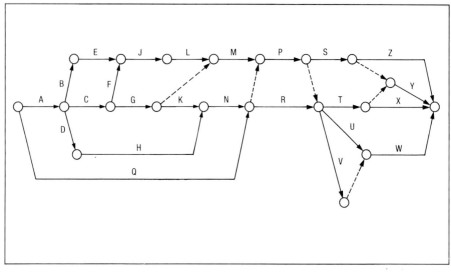

4 *One solution to exercise 2*

References
[1]British Standards Institution. *Glossary of Terms Used in Project Network Techniques*, BS4335: 1972.
[2]British Standards Institution. *Use of Network Techniques in Project Management*, BS6046: Parts 2–4: 1981.
[3]Ball, Robert, *Management Techniques and Quantitative Methods*, Heinemann, 1984.
[4]Lester, Albert, *Project Planning and Control*, Butterworth Scientific, 1982.
[5]Peters, Glen, *Project Management and Construction Control*, Construction Press, 1981.
[6]Hajek, Victor G, *Management of Engineering Projects*, McGraw-Hill, 3rd ed 1984.

Critical path analysis: 2 float and resources

In *Critical path analysis 1* the logic of the sequences of activities which go to make up a project was examined using an arrow diagram. The durations of the various activities were related by the logic and a sequence of critical activities was found which determined the project duration.

The activities which are not on the critical path have a certain amount of leeway or 'float' in their starting times which does not affect the project duration. This unit will examine float in more detail to find ways of adjusting the planned activity times to give a more efficient project.

Calculating float

To calculate the float available for a particular activity we must first identify the earliest and latest start and finish times for the activity. The earliest start time, EST, is the same as the earliest event time, EET, for its preceding event. This is the earliest time it can begin if all the activities preceding it have not caused any delays.

Thus the earliest finish time, EFT, is the EST plus the activity duration. The latest finish time, LFT, of an activity is equal to the latest event time, LET, of its succeeding event. This is the latest time it can finish without delaying the entire project. The latest start time, LST, is therefore the LFT minus the activity duration.

The difference between the earliest and latest event times is called the 'slack' of the event and is not necessarily equal to the float of the adjacent activities.

Four types of float can be identified. 'Total float', TF, is the difference between the earliest and latest finish times of the activity. It is an overall measure of the leeway available in scheduling the activity but gives no idea of the way the timing will affect subsequent activities.

'Free float', FF, is the difference between the earliest finish time and the earliest event time of the succeeding event. It is a measure of the amount the activity may be delayed without delaying subsequent activities.

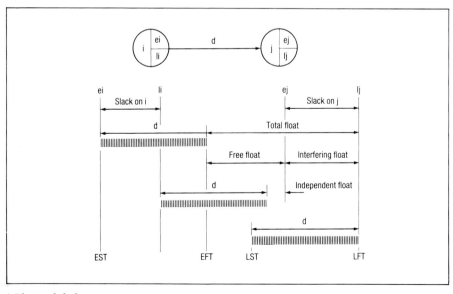

1 *Float and slack*

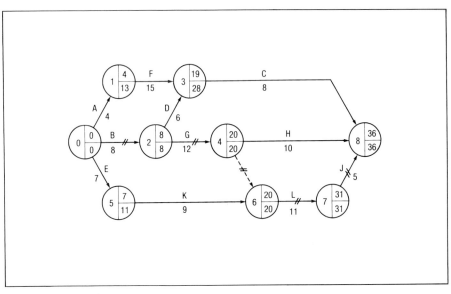

2 *Network of example*

These are the two most useful types of float.

The other two are 'interferring float', which is total float minus free float and is equal to the slack of the succeeding event, and 'independent float', which occurs when the activity begins at the LET of the preceding event and still finishes before the EET of the succeeding event. This occurs rarely and is not normally calculated. These various types of float are illustrated on a

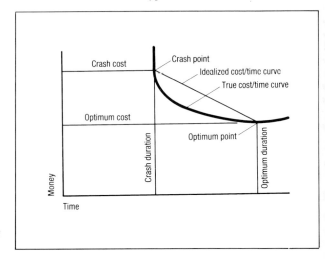

3 *Spending money to reduce the duration*

Table 1 Start and finish times and total float and free float

Title	Event Nos	D	EST	LST	EFT	LFT	TF	FF
A	0–1	4	0	9	4	13	9	0
B	0–2	8	0	0	8	8	0	0
E	0–5	7	0	4	7	11	4	0
F	1–3	15	4	13	19	28	9	0
D	2–3	6	8	22	14	28	14	5
G	2–4	12	8	8	20	20	0	0
C	3–8	8	19	28	27	36	9	9
Dummy	4–6	0	20	20	20	20	0	0
H	4–8	10	20	26	30	36	6	6
K	5–6	9	7	11	16	20	4	4
L	6–7	11	20	20	31	31	0	0
J	7–8	5	31	31	36	36	0	0

Note that all critical activities have zero total float. Other chains of activities, such as A-F-C, frequently occur which all share the same total float. This concept of float sharing is important since if early activities in the chain are delayed, the float is no longer available later on.

time-scale for a typical activity i-j in figure **1**.

Figure **2** shows a network illustrating the following logic with event times added:

■ activities A, B and E may start at the beginning of the project;
■ when B is complete, D and G may start;
■ when A is complete, F may start;
■ C is a final activity and can only start when both F and D are complete;
■ H is a final activity and can start when G is complete;
■ K follows the finish of E;
■ L cannot start until both G and K are complete; and
■ J is a final activity and can start when L is complete.

The start and finish times and the total and free float are usually tabulated as in table 1.

Exercise 1

Take the network you evolved for exercises 2 and 3 in *Critical path analysis 1* and tabulate the timings and float in the same way.

Adding cost

Suppose that a project has been planned to this stage but will take longer than desired. If money is available to spend on reducing the duration, we can now concentrate the spending where it will have most effect. The normal activity durations we have used so far are minimum-cost durations. They represent an optimum balance of time and financial resources.

By using more staff or a more expensive process the time taken can usually be improved, but there is a limit beyond which no matter how much more money you spend the time cannot be reduced or at least any reduction does not justify the expense (figure **3**). This limiting case is called the 'crash' case.

The relationship between duration and cost; between the optimum and crash positions is frequently complex but may be idealised to a straight line. The slope of this line is a cost per unit time figure (pounds per

Table 2 Assumed figures for cost reduction example

Activity	Event Nos	Optimum D	Crash D	Cost/Day	TF
A	0–1	4	3	11	9
B	0–2	8	3	7	0
E	0–5	8	5	10	4
F	1–3	15	10	8	9
D	2–3	6	5	10	14
G	2–4	12	7	11	0
C	3–8	8	5	10	9
Dummy	4–6	0	0	—	0
H	4–8	10	8	4	6
K	5–6	9	6	9	4
L	6–7	11	6	12	0
J	7–8	5	3	4	0

day) which we may apply to our network. Let us assume the figures shown in table 2.

Remember the critical path at this stage is B-G-Dummy-L-J. The project duration can only be reduced by shortening the critical activities.

Step 1   J is the cheapest critical activity and may be reduced by 2 days to crash without making others critical, ie J now takes 3 days at an extra cost of £8.
Project duration now 34 days.

Note: Total float on A, F, and C now 7 days,
   D   now 12 days,
   H   now 14 days.

Step 2   J cannot be reduced any further. B is the next cheapest critical activity but can only be reduced by 4 days before E and K become critical, ie B now takes 4 days at an extra cost of £28. Project duration now 30 days.

Note: E and K now critical.
   Total float on A, F, and C now 3 days,
   Total float on D now 8 days,

Step 3   Simultaneous reductions must be made in any parallel critical paths, ie B can only be reduced further if E is also reduced for a total cost of 6 + 7 = £13 per day. However, L is

cheaper and may reduce by 3 days before A, F, and C become critical, ie L now takes 8 days at an extra cost of £36.
Project duration now 27 days.
This may be a sufficient reduction for a total extra cost of £72. The process may be continued until a path of activities, all at crash duration, runs through the network. For this example the shortest project duration is 21 days for a total extra cost of £217, when the path E-K-L-J is at crash. Note that this was not the original critical path.

Resource scheduling

Money is a resource which can be applied in many ways, but most activities have their own particular requirements in terms of labour and equipment. We now have sufficient information to be able to adjust the timetable of a project to give maximum efficiency in the use of the available resources. If we take our example network again in its original form with optimum durations, and take the simple case where each activity requires two people to carry it out, we must first draw the network as a time-scaled bar chart, as shown in the upper part of figure 4. Note that this type of chart can only properly be drawn after the

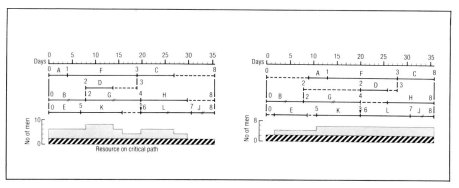

4 *Resource allocation*

5 *Rescheduling*

network has been analysed for float figures. All activities and events are shown at their earliest times. If we then add vertically for each day, the resulting graph in the lower part of figure **4** shows the pattern of staffing requirements.

This pattern is not very satisfactory with peaks and troughs. However, if we can imagine sliding the activity bars along within the limits of their float we can produce a more favourable pattern such as that shown in figure **5**, A now begins after E has finished and D fits between G and H. The resulting staff requirements build up steadily then remain constant until the project is complete.

For more complex situations, with different labour requirements on each activity, the procedure is the same but more thought is needed. If two or three types of resource are being used then several bar charts may need to be juggled at once. This may sound difficult but in practice can often be done more conveniently on paper than by computer.

Exercise 2

Using the answers to exercise 1, draw a bar chart of the project and add the labour requirements shown in table 3.

Table 3 Labour requirements

Activity	Labour	Activity	Labour
A	1	N	5
B	2	P	5
C	2	Q	6
D	2	R	4
E	4	S	4
F	3	T	3
G	4	U	3
H	3	V	2
J	4	W	3
K	2	X	3
L	4	Y	4
M	3	Z	2

Draw the labour-level graph and adjust the activity timing to keep the maximum level below nine with a smooth rise and fall at the beginning and end of the project.

Project control

Once the project is actually under way it is necessary to have a regular reporting procedure so that progress may be compared with the planned schedule. If any hold-ups occur, the network shows how much it will affect the rest of the project and where it is necessary to concentrate effort to get back to the timetable. Unless there are severe problems or the original information was in error, these changes can be made quite quickly and the new network also serves as a record of the project.

Activity-on-node networks

Where the interconnections between activities are made complex by many overlapping tasks, activity-on-node diagrams can frequently provide a more understandable picture. There are basically two types. In the simpler method of potentials each activity is represented by a box containing earliest and latest start times, the duration, and the total float (figure **6**). Links between boxes show the logic in the form, 'How long after activity A has started may B begin?'

Thus the logic shows 'start to start' links rather than the 'finish to start' links shown by the activity-on-arrow diagrams we have used so far. With the method of potentials there are no events and no dummies. The initial calculations are carried out in a similar way to CPA and immediately give activity start times and floats. Our example has been redrawn using this system in figure **7**. This diagram holds all the essential information of figure **2** and table 1.

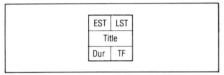

EST	LST
Title	
Dur	TF

6 *Activity box*

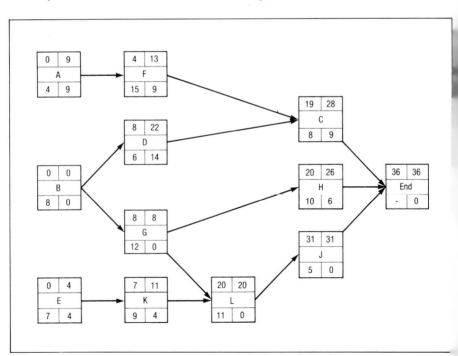

7 *Precedence diagram of example*

Exercise 3

Redraw the earlier exercise using the method of potentials.

The other activity-on-node system is the precedence network. This uses an even more informative activity box and uses logic links to show start to start, start to finish, finish to start, and finish to finish links. This is too complex to study here.

Review

These two teaching aids have given the simple basis of CPA and its associated methods. Although computer programs exist for the analytical work, experience shows that networks of at least 100 and perhaps as many as 500 activities are best constructed by hand, since the familiarity gained usually outweighs any increase in speed. The methods shown here should allow students to analyse any reasonable size project they are likely to encounter. Planning and control of the timing of the constituent activities has been covered in some detail and the basic methods of efficient allocation of costs and resource scheduling have been explained. However this has been brief and a more detailed text is recommended if these methods are to be used extensively.

Critical path analysis: 3 networks in use

The two previous articles on critical path analysis look at the theory of project planning by network analysis. In this final article the problems of applying the technique to project control will be investigated.

A technique of this sort may seem simple and logical at the planning stages, but if it cannot cope with the unexpected delays and changes of direction of a project when it is underway, it cannot be used sensibly for control purposes. That type of planning then becomes a cosmetic exercise to gain a contract instead of an integral part of project management.

Management attitudes

Most of the cases where network analysis has failed to achieve its potential can be traced back to the attitudes of managers not directly involved in compiling the network. The estimates they give for activity durations may be weighted either optimistically, to give a good immediate impression, or pessimistically, to gain praise when they finish ahead of schedule. The logic they supply may ignore delaying links or other details that seem insignificant to the ill-informed manager, but nevertheless can have a very significant effect on the overall progress of the project. This serves only to cause confusion when that aspect of the project comes to light during the construction phase.

Alternatively someone who has supplied reasonably accurate information might distrust the resulting schedule and may be reluctant to stick to it, disrupting other elements of the network. Thus it should be normal procedure to identify at an early stage of planning those managers who may not fully appreciate the advantages of CPA and take appropriate action.

Networks as policy

It is important to retain flexibility to cope with the unexpected, so networks should be regarded as guides rather than rigid structures. The act of constructing a network in a particular way, however, is a statement of policy as to the logic of the corresponding project. There may be many ways of carrying out a series of tasks but the network effectively lays down the order in which the critical tasks are to be completed.

The project manager should appreciate this aspect of the final network and ensure that any attempts to change that policy are justified and that the network is modified to show them. One of the great advantages of CPA is the ability to show the effects of policy changes on schedules, resources, and costs. It is therefore a very powerful planning tool either at the beginning of the project or part way through if changes seem necessary.

Monitoring progress

In an ideal world once the schedules have been drawn up all of the activities will take place without any delays or problems and the project will be completed on time, at the estimated cost, and using the staff and resources proposed. Where durations and logic are well defined, usually on types of projects where CPA has been used before, things do follow the plan and often achieve their targets. However, where accurate information is not available or unforeseen factors intrude, plans must be adjusted as the project progresses.

Periodically, during the project, progress should be monitored and compared with the original plan to see if adjustments should be made. The usual procedure is to make up a reporting form for each manager, which shows the activities for which they are responsible and the planned start and finish dates. Managers then fill in progress and estimates of actual finish dates. This information is then transferred to the network and any resulting changes to the schedules then passed back to the managers.

Let us take the network used in *Critical path analysis 2* and assume the project has reached day 14, also:

■ activities B, D, and E were completed at their earliest finish times;

■ activity A was delayed and has just been completed;

■ activity G will be completed in five days;

■ activity K will be completed in six days.

The original net is shown again below with event completion times added. The completed work can now be drawn as a single time elapsed arrow with adjusted times on activities in progress.

The new analysis shows that the delay on A has moved the critical path to F and C and lengthened the project by one day even though G, which was critical, will now finish a day early. If the extra day cannot be accommodated, then we can reallocate money or resources to recover the position. With network analysis this can be done.

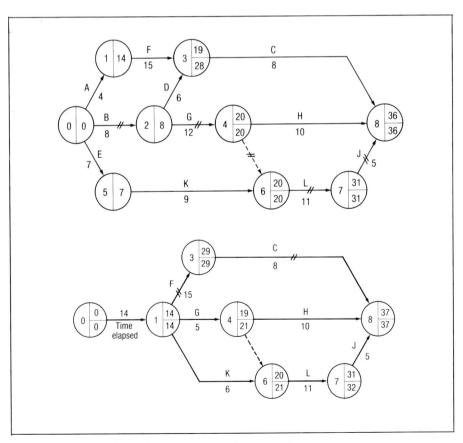

1 *Original network with actual event times and updated network*

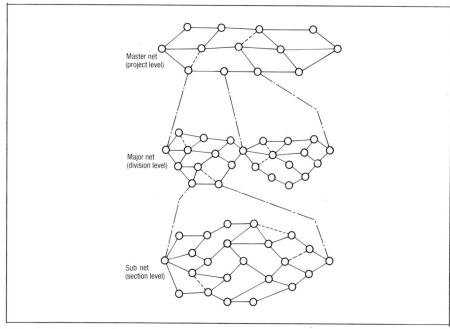

2 *Typical subnet hierarchy*

Variable logic

There are some projects where the logic
is impossible to determine beforehand,
especially when later activities depend on
the results of earlier tests or research. Many
of these cases are best illustrated by decision
trees or flow charts which fall outside our
present topic. Networks can be used to good
effect when a test procedure is used to choose
between one well-defined path and another.

For example, consider a project to
overhaul an aircraft engine. At each stage a
particular component or function is checked.
If it falls within an acceptable band of
performance, it is kept in service, perhaps
after cleaning or other routine treatment.

If it fails the test, it may be adjusted and
retested, it may be replaced, or it may be
taken apart for more detailed checks. The
logic of each stage is known but it cannot be

predetermined which path will be taken on
any particular engine.

Under these circumstances, an initial
network can be drawn to show the checking
procedure and the routine maintenance
work to be done when the engine passes
all of the tests and no remedial work is
necessary. Then after each checking activity
a separate net is drawn to show the logic
after a failed test, which continues until it
rejoins the main net.

It may be necessary to have several nets
beginning at one point or contained within
each other. (The logic is similar to the
conditional loops familiar to software
engineers.)

If statistical information is available from
past records as to the likelihood of a
particular component passing or failing a
given test, then the likelihood of following
various paths through the net can be

determined. Since the duration of each path is known from normal CPA, the distribution of times to overhaul a number of engines can be calculated. This can help considerably in planning a maintenance facility and determining the optimum level of spare parts to stock, as well as forming a very effective maintenance procedure manual.

Very large projects

When a large multi-disciplinary project is being planned, it is very convenient to break a network down into a set of smaller nets. If the project conveniently divides into separate tasks, each with only start and finish event interfacing with other sub networks, then a simple hierarchy of nets and subnets is easily constructed.

A master net shows how the separate tasks interlink and may itself be analysed as a network. Most of the arrows on that net will in turn represent another net and there may be several more layers on large projects. This has the added convenience of allowing each department, contractor, or supplier to have a sub network showing just the immediate part of the project that they are involved with and does not bog everybody down in details which are not their concern. Even very large projects then become manageable and understandable at all levels and in all departments.

Not all projects can be broken down into convenient subnets, but the benefits from such a breakdown have led to the development of ways of fragmenting large networks into manageable units. These 'frag-nets' have a number of interfaces and a full analysis can only be carried out at project level.

However, most scheduling and costing exercises can still be carried out within each department, particularly where activities are not very critical. Most larger computer programs for network analysis have facilities

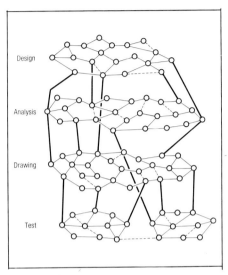

3 *Typical frag-net system*

for selecting interfaces defining fields of responsibility, or other boundary definitions, to allow parts of networks to be processed separately.

Networks as a record

After a project is completed, the first planning network and its subsequent modifications as the project proceeded provide a case history and a useful data bank for future projects. Next time a similar task or subtask is carried out this data will allow better estimates of project logic and activity durations to be made. New ideas can be tested out against detailed factual case data instead of just relying on the memories and personalities of those involved.

Problem: Durations

Activity	Time (days)	Activity	Time (days)
A	2	K	4
B	3	L	1
C	4	M	2
D	4	N	4
E	3	O	5
F	3	P	3
G	3	Q	5
H	2	R	3
J	4		

- at the beginning A, B and C may be started;
- when A is complete D and E may begin;
- after B and E are complete, J and K can start;
- F and G may start when C is complete;
- when D is complete, H can start;
- M may start only when both H and J are complete;
- L and O may start only when G has finished;
- N and Q may begin only when F, K, and L are all complete;
- P may start when M and N are complete;
- R may begin when O has been finished; and
- Q, P, and R are all final activities.

1 Set up the network, find the critical path, and tabulate the float.
2 If each activity requires two workers and only four are available on the first and last days and the total number of workers on site must never exceed eight, optimise the labour resources.
3 At the end of day 6 the following situation has arisen: C and D have just been completed; J requires another four days before completion; and K is three days away from completion.
Find the projected completion date and the new critical path by modifying the network.

Reference
BS6046 Part 1: 1984 *Use of Network Techniques in Project Management* not only gives a useful introduction to networks but a broad-ranging review of the ways projects are undertaken and the organisation of companies.

Design responsibility

It is often thought that the end of the design process is the production of a final set of drawings and other instructions which define how a product should be made, used, and maintained. Not quite. One factor remains with the designer throughout the lifetime of the product, that is, the responsibility for the design decisions.

Down in the corner of most production drawings is a series of boxes for signatures. One will be for the person who drew it and the act of signing it means that the draughtsperson accepts responsibility for the drawing as a drawing but no more. Frequently there is a checker's box. The person who signs that is saying that the individual drawing is to an approved standard and, if it is an assembly drawing, the parts can be assembled as shown. As yet no-one has said the design will work.

Many industries, particularly in technically demanding fields, have specialist engineers who analyse designs to check that their performance meets the requirements in particular ways. The most widespread of these groups are the stress specialists who check for strength, and static and dynamic stiffness. All designed objects, from computers to ships, must maintain their structural integrity in order to perform their function, hence the frequent emphasis on stress analysis.

Another frequently checked factor is weight (strictly mass), since this has significant effects on performance in a variety of ways (if a new product is too heavy, the costs of transport may be so high that it becomes uneconomic in a very competitive market). Usually these groups also have a signature box to say that the design meets their requirements.

Finally the design engineer, who most probably instructed the draughtsperson in the first place, signs the drawing affirming that it is a true representation of the required product. In doing so that designer accepts responsibility for the design.

Others may add their names approving the design before it is actually made and of course the production engineers carry the responsibility of actually making the object which the drawing represents. However, it is the designer who 'carries the can' if the product does not work through a design fault.

This is more than a passing remark. The responsibility is legally defined and, in theory at least, if a product does not perform its function or operates in an unsafe manner such that death, injury, or damage is caused, then the designer is personally responsible. This is the result of a change in the law a few years ago which has not yet been fully tested in court.

This emphasis on design responsibility makes the design engineer's task more than simply something he or she is paid to do. The responsibility is personal and if something goes wrong which can be proved to be a design flaw, the manufacturer can claim that they employ professional engineers in good faith and are not responsible for the error. The engineer's main defence should be that the design decision was based on current professional practice and was also made in good faith. If this can be shown, then the responsibility falls on the profession as a whole and not the individual.

It is therefore of paramount importance that an engineer not only belongs to a

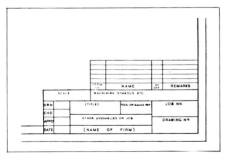

1 *The signature box on a drawing is the indicator of final responsibility*

professional body, either chartered or technician, but uses its facilities to keep up to date. Similarly there is an onus on each institution to provide the means of, or at least guidance for, updating knowledge of current practice. This is a fundamental part of the term 'professional'.

This understanding of design responsibility makes it apparent that some of the historically accepted practices in industry, where in reality a number of people may change the design, should not be allowed to happen by a responsible management. One often-repeated practice of this type arises from a split between design and production.

In many companies a 'finished' design is handed over to a production planning group who modify the design so that it may be more easily made with the company's facilities. Most of the time this causes no problems but now and again these modifications, usually made without any knowledge of the specification or the designer's intent, severely reduce or even eliminate the ability of the product to meet its required performance. Engineers with a knowledge of the company's production facilities should be included in the design team from the beginning. The aim should be to ensure that a set of design drawings can be used directly for production and that special jigs, fixtures, and tooling are allowed for in company planning.

In some companies the design function is seen as simply laying down an overall plan and some guidelines for the details. The actual details are left to the craft skills of the workforce using traditional methods or standard practice. Much building and civil engineering work follows this system. The position of every pipe or joint or course of bricks cannot be closely specified. This is quite acceptable provided the designer and the craft practitioner are aware of the responsibility involved.

Thus it may be necessary for a designer to specify details more closely where standard practice will not produce the desired results. Similarly the craft worker should be alert to situations which do not match the designer's instructions, such as specified foundations which may not be appropriate for the subsoil conditions found on excavation.

Many parts of products are designed by default, either by using standard parts or by repeating work from other areas or earlier models. Again, usually it works, but the designers should realise that they are also responsible for allowing the default to happen. It is therefore part of the designer's task to ensure that even the most routine, mundane detail will play its required part.

Design responsibility is one part of the professional environment within which an engineer operates. A better understanding of this environment should be given to students early in their course as part of the 'engineer in society' content.